Macromolecules Containing Metal and Metal-Like Elements

Volume 3

Macromolecules Containing Metal and Metal-Like Elements

Volume 3

Biomedical Applications

Edited by

Alaa S. Abd-El-Aziz
Department of Chemistry, The University of Winnipeg, Winnipeg, Manitoba, Canada

Charles E. Carraher Jr.
Department of Chemistry and Biochemistry, Florida Atlantic University, Boca Raton, Florida

Charles U. Pittman Jr.
Department of Chemistry, Mississippi State University, Mississippi State, Mississippi

John E. Sheats
Department of Chemistry, Rider University, Lawrenceville, New Jersey

Martel Zeldin
Department of Chemistry, Hobart and William Smith Colleges, Geneva, New York

A John Wiley & Sons, Inc., Publication

Chemistry Library

Published by John Wiley & Sons, Inc., Hoboken, New Jersey.
Published simultaneously in Canada.

For general information on our other products and services please contact our Customer Care
Department within the U.S. at 877-762-2974, outside the U.S. at 317-572-3993 or fax 317-572-4002.

Wiley also publishes its books in a variety of electronic formats. Some content that appears in print,
however, may not be available in electronic format.

Library of Congress Cataloging-in-Publication Data:

ISBN 0-471-66737-4
ISSN 1545-438X

Printed in the United States of America.

10 9 8 7 6 5 4 3 2 1

Contributors

Robert E. Bleicher, California State University Channel Islands, Camarillo, CA 93012

Charles E. Carraher Jr., Florida Atlantic University, Boca Raton, FL 33431 and Florida Center for Environmental Studies, Palm Beach Gardens, FL 33410 (carraher@fau.edu)

Bill M. Culbertson, College of Dentistry, The Ohio State University, Columbus, OH 43218

Minhhoa H. Dotrong, College of Dentistry, The Ohio State University, Columbus, OH 43218

Heinz-Bernhard Kraatz, Department of Chemistry, University of Saskatchewan, 110 Science Place, Saskatoon, Saskatchewan, S7N 5C9, Canada

Yitao Long, Department of Chemistry, University of Saskatchewan, 110 Science Place, Saskatoon, Saskatchewan, S7N 5C9, Canada

Eberhard W. Neuse, School of Chemistry, University of the Witwatersrand, Johannesburg, WITS 2050, South Africa (neuse@aurum.wits.ac.za)

Charles U. Pittman Jr., Mississippi State University, Mississippi State, MS 39762

Scott R. Schricker, College of Dentistry, The Ohio State University, Columbus, OH 43218

Deborah Siegmann-Louda, Florida Atlantic University, Boca Raton, FL 33431 (dlouda@fau.edu)

Mitsuhiko Shionoya, Department of Chemistry, Graduate School of Science, The University of Tokyo, Hongo, Bunkyo-ku, Tokyo 113-0033, Japan (shionoya@chem.s.u-tokyo.ac.jp)

Todd C. Sutherland, Department of Chemistry, University of Saskatchewan, 110 Science Place, Saskatoon, Saskatchewan, S7N 5C9, Canada

Contents

Preface   xi
Series Preface   xiii

1. **Organometallic Compounds in Biomedical Applications**   1
   Charles E. Carraher Jr. and Charles U. Pittman Jr.

   I. Introduction   2
   II. Case for Metal-Containing Bioactive Agents   4
       A. Tin-Containing Biocidal Polymers   5
       B. Ferrocene: A Therapeutic Role in Polymeric Systems?   6
       C. Polymeric Moderation of OsO_4 Toxicity   7
   III. Miscellaneous Polymers   7
       A. Metal Chelation Polymers   7
       B. Condensation Polymers   9
   IV. Small-Molecule Analogs   11
   V. Summary   16
   VI. References   16

2. **Metal-Labeled DNA on Surfaces**   19
   Heinz-Bernhard Kraatz, Yitao Long, and Todd C. Sutherland

   I. Introduction   20
   II. Ferrocene Nucleotides   20
   III. Ferrocene-DNA Conjugates   22
   IV. Other Metal-DNA Conjugates   34
   V. Metallated DNA   36
       A. Cu-DNA   36
       B. M-DNA   37
   VI. Summary   43
   VII. Acknowledgments   43
   VIII. References   43

3.   **Artificial DNA through Metal-Mediated Base Pairing:
     Structural Control and Discrete Metal Assembly**   45
     Mitsuhiko Shionoya

         I. Introduction   46
        II. Alternative Hydrogen-Bonding Schemes for DNA
            Base Pairing   46
       III. Non-Hydrogen-Bonding Basepairs in DNA   48
        IV. Metal-Mediated Base Pairing in DNA   49
            A. Basic Concept   49
            B. Artificial Nucleosides Designed for Metal-Mediated
               Base Pairs   49
            C. Incorporation of a Metallo-Base Pair in DNA and Its
               Effect on Thermal Stability   50
            D. Discrete Self-Assembled Metal Arrays in DNA   52
         V. Future Prospects for Artificial Metallo-DNA   54
        VI. Summary   54
       VII. References   55

4.   **Organotin Macromolecules as Anticancer Drugs**   57
     Charles E. Carraher Jr. and Deborah Siegmann-Louda

         I. General   58
        II. Anticancer Activity of Small Organotin Compounds   59
       III. Molecule-Level Studies on Monomeric Organotin
            Compounds   62
        IV. Anticancer Activity of Organotin Polymers   65
         V. Future Work   70
        VI. References   70

5.   **Organotin Oligomeric Drugs Containing the Antivirial Agent
     Acyclovir**   75
     Charles E. Carraher Jr. and Robert E. Bleicher

         I. Early History of Organotin Compounds   76
        II. Mechanisms and Reactions   76
       III. General Structures   77
        IV. Acyclovir   80
         V. Bioactivity of Related Compounds   81
        VI. Experimental Work   82
       VII. Results and Discussion   83
      VIII. References   86

6.   **Polymeric Ferrocene Conjugates as Antiproliferative Agents**   89
     Eberhard W. Neuse

         I. Introduction   90
        II. The Ferrocene–Ferricenium System in the Biological
            Environment   92

III. Polymer–Drug Conjugation as a Pharmaceutical Tool for
Drug Delivery ... 98
IV. Polymer–Ferrocene Conjugates: Synthesis and Structure ... 100
    A. The Carrier Component: Structural Considerations ... 101
    B. Conjugates of Amide-Linked Ferrocene ... 102
    C. Conjugates of Ester-Linked Ferrocene ... 109
V. Bioactivity Screening ... 110
VI. Summary and Conclusions ... 113
VII. Acknowledgments ... 115
VIII. References ... 115

**7. Polymeric Platinum-Containing Drugs in the Treatment
of Cancer ... 119**
Deborah W. Siegmann-Louda and Charles E. Carraher Jr.

    I. Introduction ... 120
    II. Basic Mechanisms of Pt(II) Complex Formation ... 121
    III. Nomenclature ... 125
    IV. Currently Approved Platinum-Containing Compounds ... 125
    V. Properties of Cisplatin ... 127
    VI. Structure–Activity Relationships ... 130
    VII. Polymer–Drug Conjugation Strategy and Possible
    Benefits ... 133
        A. Polymers as Carriers ... 134
        B. Polymers as Drugs ... 135
        C. General ... 136
    VIII. Mainchain-Incorporated *cis*-Diamine-Coordinated
    Platinum ... 137
        A. Simple Amine Derivatives ... 137
        B. Amino Acid Derivatives ... 141
        C. Other Nitrogen–Platinum Products ... 143
        D. Solution Stability ... 144
        E. Thermal Stability ... 145
        F. Antiviral Activity ... 145
    IX. Platinum Carrier-Bound Complexes via Nitrogen
    Donor Ligands ... 147
        A. Pt-Polyphosphazenes ... 147
        B. Slowly Biofissionable Pt–N Complexes Anchored
           through Primary and Secondary Amines ... 149
        C. Biofissionable Pt–N Complexes Anchored through
           Primary and Secondary Amines ... 154
    X. Pt–O-Bound Polymers ... 161
    XI. Mixed Pt–O/Pt–N-Bound Polymers ... 180
    XII. Future Work ... 182
    XIII. Acknowledgments ... 184
    XIV. References ... 185

8. New Organic Polyacid–Inorganic Compounds for Improved Dental Materials   **193**

Bill M. Culbertson, Minhhoa H. Dotrong, and Scott R. Schricker

    I. Introduction   194
   II. Glass Ionomer Technology   194
      A. Amino Acid–Modified Glass Ionomers   198
      B. *N*-Vinylpyrrolidone (NVP)-Modified Glass Ionomers   201
  III. New NVP-Modified Glass Ionomers: Experimental Work   202
      A. Materials   202
      B. Polymer Synthesis   203
      C. Characterization   203
      D. Physical Properties   203
   IV. Results and Discussion   204
    V. Conclusions   205
   VI. References   206

Index   **209**

Preface

Metals are essential elements to life. Metal-containing biomedical polymers are all about us and metalloenzymes serve as basic building blocks of life in both animals and plants. Metal-containing polymers serve as structural material in our bones and teeth and the shells of sea creatures. As metal ion polymer combinations they serve to determine transmission across membranes, transmission of signals to the brain, etc., but they are also critical components of many biopolymers that allow life to occur. The appearance of metal-containing macromolecules in the human body is extensive and includes such metals as iron (transferrin, hemoglobin), molybdenum (xanthine oxidase), vanadium (hemovanadin), zinc (carbonic anhydrase), and copper (hepatocuprein). Frequently, metal atoms serve as redox centers in enzymes and depots for metal storage. Therefore, synthetic metal-containing polymers might function as therapeutic agents through their ability to reduce oxidizing agents or oxidize reductants or provide sources for crucial metal uptake by biopolymers.

The use of organometallic medicinals is also widespread. Just as small molecules such as cisplatin, the most widely used anticancer drug, are important agents to fight disease, polymeric analogs have been synthesized that exhibit greater specificity, lower toxicities, and increased activity because of their polymeric nature.

The use of metal and metalloid-containing macromolecules is widespread. For example, polysiloxanes are used as biomaterials rather than drugs. Polysiloxanes are widely used as contact lens and in the reconstruction of finger, hip, toe, and wrist joints and in the manufacture of artificial lungs, skin, dialysis units, orbital floors, tracheal stents, brain membranes, ear frames, and hearts.

Metal-containing polymers can act by several mechanisms. For example, the remainder of the polymer might modify the metal site activity or it may simply hold or carry the metal sites. The polymer may simply act within a controlled release mechanism. For instance, metal ions can coordinate well to Lewis base centers. When surrounded by a ligand of defined structure, the remaining coordination sites have a specified geometric relationship imposed. Thus, the Zn(II) complex of xylylbicyclam is a potent anti-HIV agent because it recognizes and binds to the coreceptor used by HIV for membrane fusion and cell entry. This specific site coordination, coupled with the shape and flexibility of the ligand, allows for induced-fit recognition of this specific biological target. Extension of this concept through the use of metal-containing macromolecules is an area ripe for future investigation.

Cancer is a multiplicity of diseases. Many aspects of chemical treatment can be, and are being, applied to treat cancers. A drug can be bound to a polymer and either released to the body or be active in the bound form. Several chapters in this volume concentrate on the most up-to-date research that deals with the treatment of cancer employing metal-containing polymers. Chapters also deal with antiviral uses of metal-containing macromolecules. Metal-containing biomaterials are also important in areas in addition to their use as drugs. For example, they serve as scaffolds or building materials in our teeth and bones, etc. A chapter illustrating the use of metal-containing macromolecules in dentistry and dental uses of metal-containing materials either as drugs or as speciality materials is present in this volume. The application of metal-containing dendritic structures for biomedical uses and the use of artificial metallo-DNAs is also included in this volume.

Basic and applied information that allows the reader to understand and follow progress within these important areas has been incorporated into the individual chapters throughout this volume.

Synthetic metal-containing macromolecules are important members of the biomaterials community and their importance will continue to grow.

Alaa S. Abd-El-Aziz
Charles E. Carraher Jr.
Charles U. Pittman Jr.
John E. Sheats
Martel Zeldin

Series Preface

Most traditional macromolecules deal with less than 10 elements (mainly C, H, N, O, S, P, Cl, F), whereas metal and semi-metal-containing polymers allow properties that can be gained through the inclusion of nearly 100 additional elements. Macromolecules containing metal and metal-like elements are widespread in nature with metalloenzymes supplying a number of essential physiological functions including respiration, photosynthesis, energy transfer, and metal ion storage.

Polysiloxanes (silicones) are one of the most studied classes of polymers. They exhibit a variety of useful properties not common to non-metal-containing macromolecules. They are characterized by combinations of chemical, mechanical, electrical, and other properties that, when taken together, are not found in any other commercially available class of materials. The initial footprints on the moon were made by polysiloxanes. Polysiloxanes are currently sold as high-performance caulks, lubricants, antifoaming agents, window gaskets, O-rings, contact lens, and numerous and variable human biological implants and prosthetics, to mention just a few of their applications.

The variety of macromolecules containing metal and metal-like elements is extremely large, not only because of the larger number of metallic and metalloid elements, but also because of the diversity of available oxidation states, the use of combinations of different metals, the ability to include a plethora of organic moieties, and so on. The appearance of new macromolecules containing metal and metal-like elements has been enormous since the early 1950s, with the number increasing explosively since the early 1990s. These new macromolecules represent marriages among many disciplines, including chemistry, biochemistry, materials science, engineering, biomedical science, and physics. These materials also form bridges between ceramics, organic, inorganic, natural and synthetic, alloys, and metallic materials. As a result, new materials with specially designated properties have been made as composites, single- and multiple-site catalysts, biologically active/inert materials, smart materials, nanomaterials, and materials with superior conducting, nonlinear optical, tensile strength, flame retardant, chemical inertness, superior solvent resistance, thermal stability, solvent resistant, and other properties.

There also exist a variety of syntheses, stabilities, and characteristics, which are unique to each particular material. Further, macromolecules containing metal and metal-like elements can be produced in a variety of geometries, including linear, two-dimensional, three-dimensional, dendritic, and star arrays.

In this book series, macromolecules containing metal and metal-like elements will be defined as large structures where the metal and metalloid atoms are (largely) covalently bonded into the macromolecular network within or pendant to the polymer backbone. This includes various coordination polymers where combinations of ionic, sigma-, and pi-bonding interactions are present. Organometallic macromolecules are materials that contain both organic and metal components. For the purposes of this series, we will define metal-like elements to include both the metalloids as well as materials that are metal-like in at least one important physical characteristic such as electrical conductance. Thus the term includes macromolecules containing boron, silicon, germanium, arsenic, and antimony as well as materials such as poly(sulfur nitride), conducting carbon nanotubes, polyphosphazenes, and polyacetylenes.

The metal and metalloid-containing macromolecules that are covered in this series will be essential materials for the twenty-first century. The first volume is an overview of the discovery and development of these substances. Succeeding volumes will focus on thematic reviews of areas included within the scope of metallic and metalloid-containing macromolecules.

Alaa S. Abd-El-Aziz
Charles E. Carraher Jr.
Charles U. Pittman Jr.
John E. Sheats
Martel Zeldin

CHAPTER 1

Organometallic Compounds in Biomedical Applications

Charles E. Carraher Jr.

Florida Atlantic University, Boca Raton, Florida and Florida Center for Environmental Studies, Palm Beach Gardens, Florida

Charles U. Pittman Jr.

Mississippi State University, Mississippi State, Mississippi

CONTENTS

I. INTRODUCTION	2
II. CASE FOR METAL-CONTAINING BIOACTIVE AGENTS	4
A. Tin-Containing Biocidal Polymers	5
B. Ferrocene: A Therapeutic Role in Polymeric Systems?	6
C. Polymeric Moderation of OsO_4 Toxicity	7
III. MISCELLANEOUS POLYMERS	7
A. Metal Chelation Polymers	7
B. Condensation Polymers	9
IV. SMALL-MOLECULE ANALOGS	11
V. SUMMARY	16
VI. REFERENCES	16

Macromolecules Containing Metal and Metal-Like Elements,
Volume 3: Biomedical Applications, edited by Alaa S. Abd-El-Aziz,
Charles E. Carraher Jr., Charles U. Pittman Jr., John E. Sheats, and Martel Zeldin
ISBN: 0-471-66737-4   Copyright © 2004 John Wiley & Sons, Inc.

I. INTRODUCTION

The toxicity as well as therapeutic value of organometallics is well known. The introduction of metal ions within biological macromolecules such as proteins and nucleic acids is a continuing area of research. The appearance of metal-containing macromolecules in the human body is extensive and includes such metals as iron (transferrin, hemoglobin), molybdenum (xanthine oxidase), vanadium (hemovanadin), zinc (carbonic anhydrase), and copper (hepatocuprein). The use of organometallic medicinals is widespread. Some examples include merbromine (mercurochrome), meralein (mercury, antiseptic), silver sulfadizine (prophylactic treatment for severe burns), arsphenamine (arsenic, antimalarial), 4-ureidophenylarsonic acid (therapeutic for ameblasis, tryparsamide, and Gambian sleeping sickness), and antimony dimercaptosuccinate (schistosome). Table 1 contains a brief listing of additional metal-containing drugs.

Perhaps the earliest written record of the use of medicines is the "Ebers" papyrus, from about 1500 BC. This describes more than 800 "recipes," some of which contain substances today known to be toxic, including hemlock, aconite

Table 1 General Biological Uses for Metal-Containing Drugs

Metal	Medical Use
Au	Arthritis, gout
Ag	Antiseptic agent, prophylacetate
As, Sb	Bactericides
Bi	Skin injuries, diarrhea, alimentary diseases
Co	Vitamine B_{12}
Cu	Algicide, fungicide, insecticide
Ga	Antitumor agent
Hg	Antiseptic
Li	Manic depressoin
Mn	Fungicide, Parkinson's disease
Os	Antiarthritis
Pt, Pd	Antitumor agents
Rb	Substitute for K in muscular dystrophy; protective agent against adverse effects of heart drugs
Ru, Rh, Os	Experimental antitumor agents
Sn	Bactericide, fungicide
Ta, Si	Inert medical applications as gauzes, implants
Zn	Fungicide

(ancient Chinese arrow poison), opium, and metals, including lead, copper, and antimony.[1] Use of mercury in medicine in ancient Greece was described by Dioscorides, and by the Persian Ibn Sina of Avicenna (980–1036 AD), for use against lice and scabies. He also reported observations of chronic mercury toxicity.[2]

Arsenic is another metal known to the ancients with toxic as well as medicinal properties. The sulfides of arsenic, which were roasted, were described by Discorides in the first century AD as medicines as well as colors for artists. There is evidence that arsenic was used as a poison in Roman times.[3] Medieval alchemists were well aware of the poisonous nature of arsenic compounds, which were used in various recipes. Paracelsus, the Swiss physician, used arsenic compounds as medicinal agents.[4] Arsenic was widely used as a pesticide in the form of calcium arsenate following the turn of the 20th century.

Paracelsus understood the relationship between medicines and poisons as stated in the third of his 1536 Sieben Defensiones:[2] "What is not a poison? All things are poisons and nothing is without toxicity. Only the dose permits anything not to be poisonous. For example, every food and every drink is a poison if consumed in more than the usual amount; which proves the point." We might add the mode of delivery is also critical. For instance, water injected down the windpipe kills.

Metal-containing compounds which are toxic to fungi, bacteria, protozoa and other disease-causing agents can often be toxic to humans. Mercury toxicity has been known since ancient times. Romans, for example, used criminals in the cinnabar mines in Spain as the life expentancy of a miner was just 3 years.[5] Other examples of metals that were mined for a variety of uses include lead and arsenic. Egyptians were known to use lead back to 5000 BC, and mines in Spain date to about 2000 BC.[6] It has been suggested that the use of lead in making wine and other products, and the use of lead in pipes could have contributed to widespread lead poisoning in Roman times.[7]

Evidence for lead contamination in the environment from anthropogenic activities including mining, smelting, and combustion is historically preserved in various sites such as ice in Greenland, lake sediments and peat in Sweden, and elsewhere. Studies of such historical samples have shown, for example, that the $^{206}Pb/^{207}Pb$ ratios of about 1.17 in 2000 year-old corings are similar to ratios of the lead sulfide mined by the Greeks and Romans during this time period. By contrast, local northern sources such as those in Sweden that are uncontaminated by long-range atmospheric transport, exhibit a much higher ratio (about 1.53). In fact, these studies confirm that lead use and deposition greatly decreased following the decline of the Roman Empire.[8]

Although lead has been used since ancient times for medicinal purposes, its toxic properties were also understood. Thus, lead colic was reported by Hippocrates, and in about 50 AD, poisoning of lead workers was documented by Pliny. Ramazzini observed toxicity to potters working with lead in 1700 AD, but it was not until 1933 that Kehoe demonstrated wide exposure to lead in the environment. Lead produces adverse effects on children with respect to behavior and reduced IQ scores, even at very low levels.[7]

The use of metal and metalloid-containing polymers in biomedical applications is widespread. Polysiloxanes are the primary materials currently employed. Organotin polymers are also widely used and polyphosphazenes have been used. Chelated metals are widely employed in dental applications. Many others are waiting in the wings for use in biomedical applications including additional organotin polymers, ferrocene-containing polymers, as well as platinum polymers. Many more may emerge in the near future as signaled by the activity of small molecules containing metals as both drugs or essential sites for the biomedical activity. This brief chapter is intended to suggest potential areas where metal-containing macromolecules may play a significant role based on the study of such small metal-containing molecules. The use of small metal-containing molecules in biomedical applications was highlighted in a Chemical and Engineering News article called "The Bio Side of Organometallics," which reviewed a major meeting on this topic held in July 18–20, 2002, at ENSCP where 120 participants from 25 countries discussed this topic.

Areas covered in the other chapters of this volume will not be dealt with here. Instead, additional efforts will be described.

II. CASE FOR METAL-CONTAINING BIOACTIVE AGENTS

We have seen that metal- and metalloid-containing compounds exhibit a wide range of biological and biocidal activities, some of which have been employed in medicines and drugs. Polymers containing metal or metalloid functions become a natural extension of this effort. Just as organic compound drugs have been chemically bound to polymers or physically imbibed into polymer matrices in order to provide a variety of useful advantages, the same opportunities exists for using metal and metalloid species. The use of polymeric drugs provides many possible advantages; a few of them are described here:

1. Controlled release of the active agent either by diffusion from a matrix or hydrolytic/enzymatic cleavage from the polymer carrier. This allows a sustained and more steady delivery of the active agent within the body or from patches applied externally.
2. External application of a medicine for transdermal diffusion is possible for more volatile or water-soluble compounds.
3. The polymer can be tailored to modify the solubility of the active medicinal agent. Its hydrophobicity/hydrophilicity is dominated by the polymer carrier.
4. Proper design of molecular weight can greatly reduce the excretion rate, thereby increasing efficacy and reducing the required dosage. Control of chain length can also be employed to "isolate" or prevent movement of the polymer drug past barriers in the body such as the blood–brain barrier.
5. The polymer can enhance the concentration of the active agent in specific tissues or locations. Polymers containing attached specific binding recognition agents (e.g., antibodies, hormones) can bind to biological receptor sites. If that polymer

also has medicinal molecules attached, these active agents are now delivered to and concentrated in the targeted locations. This highly specific delivery to cells with the targeted receptors is but one end of the spectrum. Molecular weight and hydrophobic-hydrophilic partitioning effects can also be employed. Further, the large size of the polymer often allows preferential attachment to cell walls, again effectively increasing the delivery time and lowering toxic effects related to the kidney and renal system.

A. Tin-Containing Biocidal Polymers

Perhaps the best known use of metal-containing polymers to deliver toxic substances at controlled rates are the polymeric tin methacrylates that have been used extensively in antifouling marine coatings.[9] Organotins have had widespread application in biocidal compositions.[10] Trialkyltin derivatives are the most effective toxins against marine organisms that foul ship bottoms, ultimately leading to barnacle growth as the climax organisms after a complex multiorganism colonization process.[11] Specifically, tributyltin gave the best antifouling activity when readily hydrolyzed from a carboxylate function.[12,13] Therefore, a large number of multilayer coating systems were developed where tributyltin methacrylate monomer units were present that underwent hydrolysis, initially producing tributyltin hydroxide.[9,14] The released tin moieties killed or inhibited the organism succession needed to colonize the ship bottom surfaces, reducing fouling. Related systems containing triphenyltin, triphenyllead, tributyllead, organoarsenic compounds, and copper(I) oxide have been studied and used.

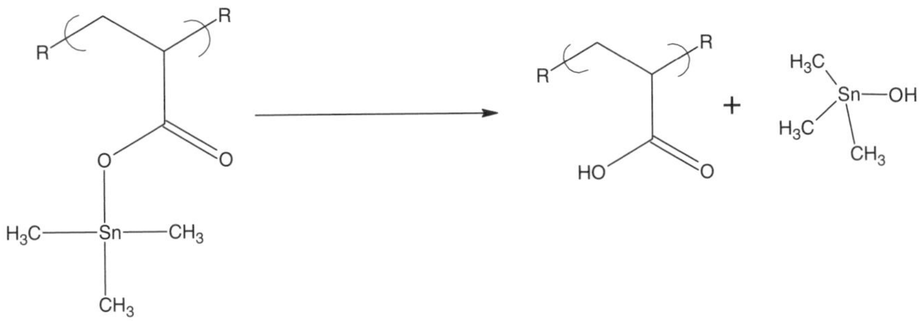

1   **2**   **3**

The use of biocidal antifouling coatings to store a toxic agent in large quantity and then deliver it more slowly at a location in which the toxicant is needed, suggests that polymers containing metals and metalloid functions, should become a fertile field for drug/pharmaceutical development and agricultural chemicals. An active agent might be released from the polymer or might function while in the polymeric form. Therefore, in the next example, we offer a rational for how the first synthetic organometallic compound prepared and characterized, ferrocene,[15–17] might serve as a therapeutic agent when it is present in a polymer. In this case, however, a toxic agent would not be released. Instead, the iron center in ferrocene could function as a redox center to reduce the concentration of oxidants that cause cell damage via free-radical chemistry.

B. Ferrocene: A Therapeutic Role in Polymeric Systems?

The biochemistry of cancers is enormously complex, but it has generally been accepted[18-20] that free radicals play many important roles, notably in carcinogenesis and tumor promotion. Superoxide anion radical, hydrogen peroxide, and hydroxyl radical generation are formed during O_2 reduction in respiring cells. Protection is afforded by enzymes, such as superoxide dismutase, which converts superoxide ion into hydrogen peroxide and O_2. Catalase and glutathione peroxidase eliminate hydrogen peroxide, preventing Fenton reactions, which generate hydroxyl radicals.[21,22] Superoxide-generating promoters and reductions in superoxide dismuturase activity lead to increased superoxide levels in transformed cells.[22] Ultimately, radical species react at nuclear DNA causing replication errors during mitosis which is a causative factor in malignancy. Carcinogenic and metastatic processes are inhibited by antioxidants and free-radical scavengers.[18,22,23] Carcinostatic activity is exhibited by some metal complexes that serve as scavengers for superoxide.[19,24] Free-radical scavenger activity can be the operating mode of some anticancer drugs.[25]

Many metal-containing compounds are good reducing agents. For example, ferrocene is readily oxidized to stable ferrocenium by hydroxyl radicals or other radicals. These radicals are reduced (e.g., $^{\bullet}OH \rightarrow {}^{-}OH$). Furthermore, ferricenium is reduced by superoxide. Therefore, this redox activity can serve to remove both of these deleterious oxidants. This suggests that ferrocene might serve as a useful therapeutic center if its solubility, pharmocokinetic response, distribution in the body, and other required features could be suitably tailored. The use of polymer-bound ferocene species could serve to tailor these requisite properties. Thus, water-soluble ferrocene-containing polymer conjugates have been synthesized for this purpose,[26-28] and water-soluble monomeric ferricenium salts were found to have powerful antineoplastic activity, curing cells under the Ehrlich ascites tumor testing protocol.[27,28] It seems likely that metal-containing polymers should be a rich source of potential future therapeutic agents.

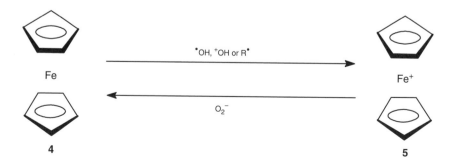

C. Polymeric Moderation of OsO$_4$ Toxicity

Polymer-binding of highly toxic metal reagents can ameliorate the toxicity of potentially beneficial osmium compounds. Osmium carbohydrate polymers, called *osmarins,* have been synthesized[29] and proposed for use as active agents to treat arthritis.[30] This proposal was based on the use of osmium tetroxide for about 50 years to treat human arthritis, mostly in Europe.[31] This treatment is controversial because of the high toxicity of OsO$_4$. Solutions of OsO$_4$ are injected into the synovial space of diseased joints. The treatment is longlasting when successful according to Swiss scientists who examined 73 successfully treated patients.[32] Osmium-containing material remained in the joints for as long as 5 years, suggesting a long-term biological effect.[32] The use of the carbohydrate osmium polymers for this purpose was proposed[30] to produce the long-term beneficial effects of osmium deposits while avoiding trauma associated by unselective OsO$_4$ oxidations encountered after OsO$_4$ injections. The osmarin polymers are prepared by reacting Os(OAc)$_4$ with glucose.[29] Polydisperse, polyanionic spherical polymers are obtained with molecular weights from a few thousand to a few hundreds of thousands. They are thought to be carbohydrate-solubilized OsO$_2$ species. Unlike the highly toxic OsO$_4$, the osmarin polymers have low acute toxicities and does of 1 g/kg have been administered to mice with no mortality.[33] When injected into pig tarsal joints, these polymers bind to all of the tissue surfaces facing the synovial space but cause far less trauma than that observed by injecting OsO$_4$.

The tin, iron, and osmium examples discussed above allow one to imagine all sorts of possibilities for metal- and metalloid-containing polymers with biomedical applications. In the next section a series of selected biomedical applications of metal-bound polymers is reviewed. This is followed by a short review of some representative small molecules containing metals which have biomedical uses.

III. MISCELLANEOUS POLYMERS

A. Metal Chelation Polymers

As noted in Table 1, copper-containing materials have been utilized as algicides, fungicides, and insecticides. Numerous copper-containing polymers have been formed through the chelation of copper. Donaruma and coworkers described the use of poly(thiosemicarbazide) copper(II) complexes, **6**, as algicides and molluscicides.[34] They showed that this polymer released copper(II) slowly and demonstrated that the product could be used to construct reusable cartridges for schistosomaisis and algae control. Various R$_1$ groups were studied.

Numerous iron-containing polymers have been synthesized and studied for biomedical application. Here we will look at simple iron(II) chelating polymers developed for medical applications.

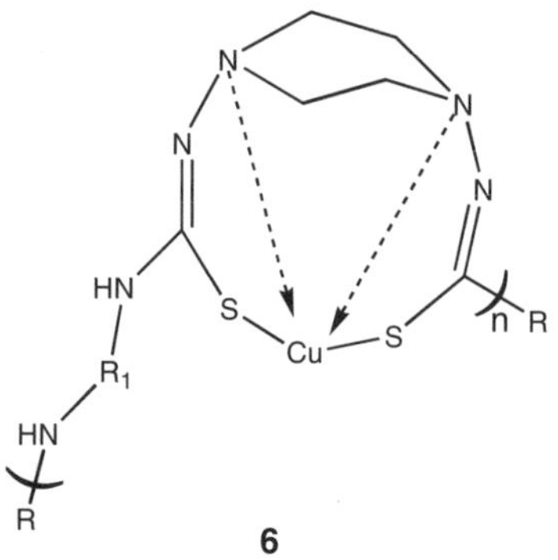

6

Medical interest in developing iron chelating drugs is due almost entirely to the potential use in removing iron from patients experiencing iron overload.[35–39] There are two major causes of iron overload. They include the over consumption and under removal of iron. The under removal of iron is generally encountered in a disease called β-*thalassemia*, better known as *Cooley's anemia*. Iron poisoning is a problem in small children caused by the inadvertent ingestion of iron-containing materials. Before the use of iron chelation therapy in the 1960s, such conditions were generally fatal. Today good treatments exist involving the use of powerful iron chelating agents. Cooley's anemia is a genetic disorder, rare in the United States but widespread in the Mediterranean area, the Middle East, India, and Southeast Asia. The disease is characterized by an inability to synthesize adequate amounts of the beta chain of hemoglobin. Because excess alpha chains cannot form soluble tetramers, precipitation occurs in red cell percursors leading to their death and to the anemia.

A number of naturally occurring iron chelators exist including desferrioxamine B, enterobactin, and ferrichrome.[40] Each of these bind the iron through various chelating groups, generally oxygen and nitrogen. Because of its low toxicity and high ability to chelate iron, desferrioxamine B (DFO), is now available for commercial use under the tradename of Desferal.[41] While DFO is effective in removing large amounts of iron rapidly, it has a short plasma residence time.

Research in the chelation of iron is sponsored by various Cooley's Anemia Foundations, the National Institute of Arthritis, Diabetes, and Digestive, and Kidney Diseases of the National Institutes of Health. Much of effort has been directly toward mimicking the naturally occurring siderophores such as desferrioxamine by inclusion of hydroxamic acids, catechols, and phenols into a variety of structures. This gives compounds that have vary high stability constants for iron but are nontoxic and active in the human body.

Winston and coworkers did extensive work in this area to develop polymer-containing materials that bind iron.[35–39] These studies illustrate the formation of metal-containing materials, not as a drug itself, but to study the effectiveness of the chelating agent as a potential drug for biomedical use.

Hydroxamic acid is the functional group responsible for binding the iron in DFO. It has been known since 1869.[40] The first hydroxamic acid polymer was made in 1946. A number of hydroxamic acid polymers with controlled spacing have been developed that effectively bind iron with iron binding log K values around 30. The structure of one of these is given below, where R_1 is the spacer.

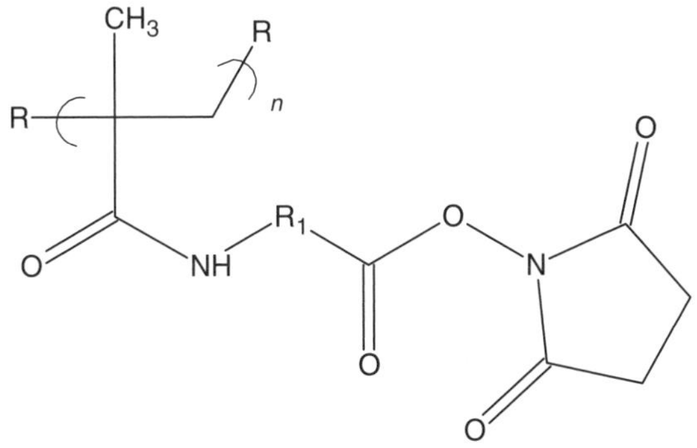

7

B. Condensation Polymers

A number of metal-containing polymers have been touted as potential anti-cancer agents. This volume contains reviews that describe the most common of these: platinum-, tin-, and ferrocene-containing polymers. However, other metal-containing polymers are also known. Norihiro[42] described the use of oligomeric organogermanium compounds with the following formula, where R and R_1 are H or lower alkyl and n is equal to or greater than 2. While the monomer is not active, the oligomer inhibits experimental tumor growths at dose levels of 50–200 mg/kg.

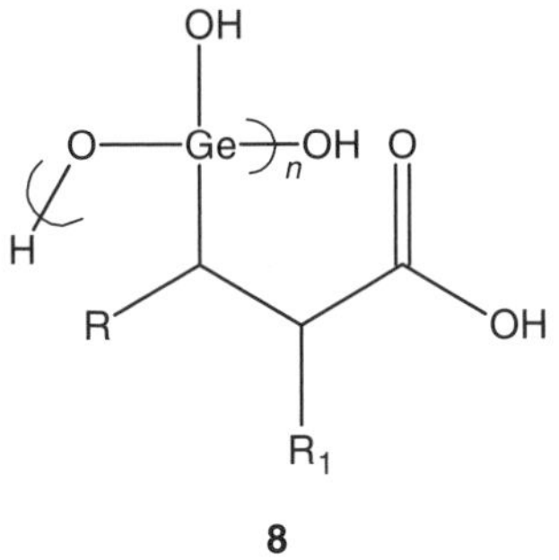

8

Carraher and coworkers[43] have synthesized a number of group VA-containing polymers that exhibit both anti-tumor and anti-bacterial actively. Polymers formed from reaction with triphenylantimony dichloride and thiopyrimidine exhibit good inhibition of Balb/3T3 cells at concentrations down to 10 µg/mL.

Products from triphenylantimony dichloride and cephalexin show good inhibition of Balb/3T3 cells to 2 μg/mL; polymers from triphenylarsenic dibromide and cephalexin exhibit cells to 50 μg/mL; and those from trimethylantimony dibromide and cephalexin show good inhibition to 15 μg/mL.[44] The structure for the triphenylantimony dichloride and cephalexin product is given below.

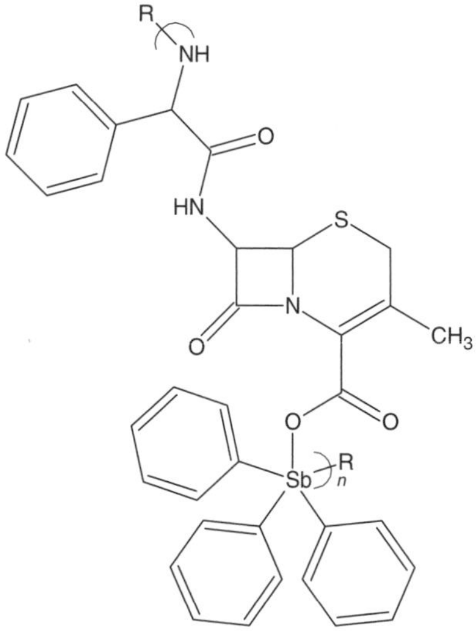

9

A number of antimony(V) polyamines were synthesized and biologically characterized as antibacterial agents.[45] The products inhibited a wide range of bacteria, including *Actinobacter calcoaceticus, Alcaligenes faecalis, Branhamella catarrhalis, Enterobacter aerogenes, Escherichia coli, Klebsiella pneumoniae, Neisseria mucosa, Psedomonas aeruginosa, Staphylococcus epidermis,* and *Staphylococcus aureaus.* These materials also effectively inhibited HeLa cells at concentrations of about 5 μg/mL. A sample structure is given below.[45]

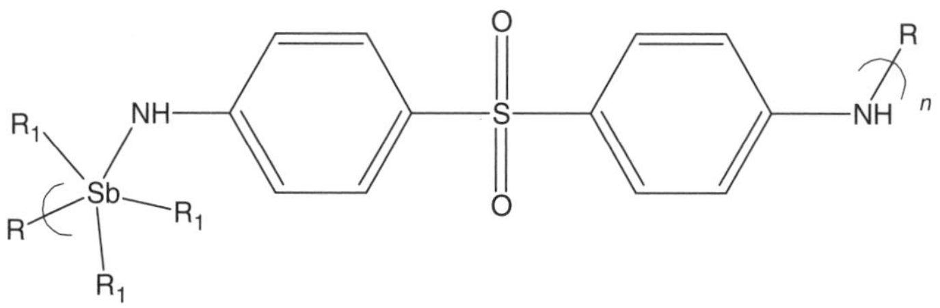

10

IV. SMALL-MOLECULE ANALOGS

There are many examples of current efforts where "small" organometallic compounds are being investigated for biomedical application. Following is a short description of some of these.

As shown in the present volume, metal-containing polymeric materials are being widely considered as critical agents in the war against cancer. In a similar manner, small molecule compounds are being investigated in the fight against cancer.

Jaouen, Top, Vessieres, and coworkers are utilizing organometallic compounds for treating breast cancer, the most common cancer among women, affecting about one in eight females.[46] The most widely employed drug employed in the treatment of breast cancer is tamoxifen.[47,48] While generally well tolerated, it has some drawbacks. After extensive use, resistance to the drug can develop. It also increases the risks of uterine cancer, blood clotting in the lungs, and is not as effective against hormone-independent tumors that account for about one-third of breast cancers.

The main function of tamoxifen is to block or interfere with the action of estradiol, one of a group of female hormones known as esterogens. Tamoxifen metabolizes forming hydroxytamoxifen that binds to an estrogen receptor preventing the cell from growing and dividing.

Jaouen and coworkers are investigating a number of cyclopentadienyl, Cp, metal complexes for tamoxifen-like activity. They initially focused on titanocene dichloride, Cp_2TiCl_2. While early results showed promise, the tamoxifen-titanocene derivative acted like an estrogen, promoting the growth of breast tumors rather than preventing their growth. The structure of one of these derivatives is given below.

In a similar study, ferrocene-substituted tamoxifens, "ferrocifens," exhibited good antiproliferative effects on several breast cancer cell lines.[49] These ferrocifens are the first drugs that show good activity against both the hormone-dependent and hormone-independent breast cancer cells.

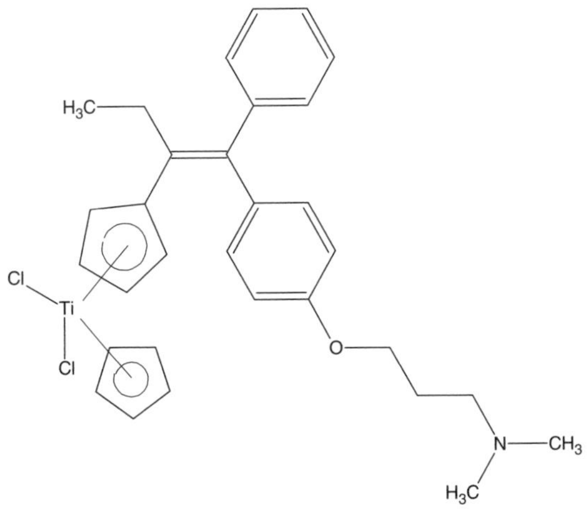

The hydroxyl derivative, hydroxyferrocifen, shown below, also inhibits cell growth in kidney cancer and ovarian cancer cell lines. Live animal tests in rats and mice show that some of the hydroferrocifen derivatives are less toxic than tamoxifen itself.

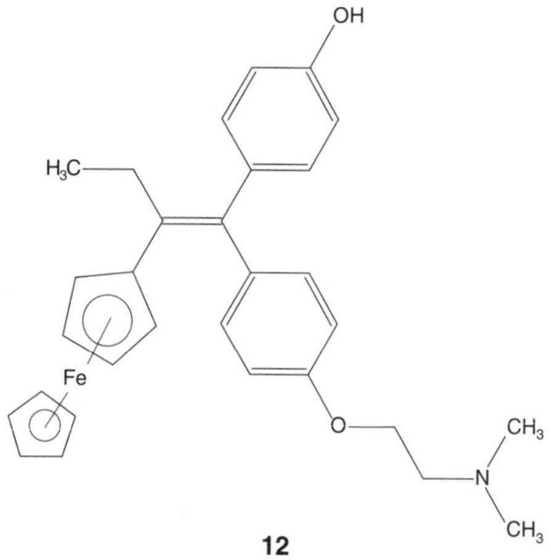

12

This group also synthesized and studied a series of Ru, W, and Co complexes of 17-estradiol.[50] These modified hormones showed good affinity for the binding at the estrogen receptor site. One of these compounds is illustrated below.

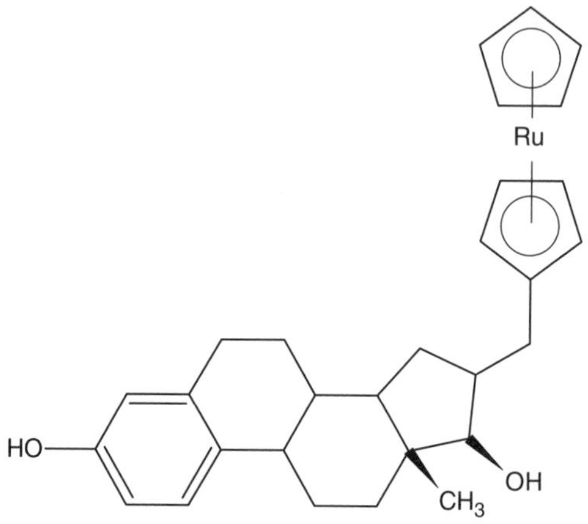

13

Oscella, Rosenberg, Hardcastle, and coworkers have been active in developing a number of metal-containing drugs containing such metals as Co, V, Ru, Pt, and Fe.[51] They have also been developing Os-containing drugs for the purpose of specifically targeting telomerase. Telomerase is known to be involved in the growth of at least some cancers. For each cell cycle, the telomers that appear at the ends of chromosomes are shortened to ensure that after 50–70 replication cycles the cell ceases to divide and succumbs to programmed cell death or apoptosis. Telomerase allows tumor cells to circumvent this normal process by elongating the length of the telomer ends. In fact, one reason for the efficacy of cisplatin against testicular cancer is believed to be its ability to inhibit telomerase. Unfortunately, cisplatin is not specific to telomerase inhibition, which brings about the occurrence of undesired side effects.

It is important that the non-tumor-specific toxicity be as mild as possible. In search of such a drug, this research group settled on one class of osmium derivatives. These compounds have a nucleobase-like ligand and a phosphine or phosphite ligand capable of imparting water solubility. Both of these components are coordinated to a massive but chemically inert triosmium–nonacarbonylhydrido core. One of these structures is given below. These compounds differ from cisplatin since they do not alkylate DNA. Instead they interfere with the catalytic activity of the telomerase through their steric bulk. Some of these compounds exhibited good telomerase inhibition with little evidence of nonspecific cytotoxicity, suggesting that they interact directly with the enzyme. After several replication cycles, the cells died.

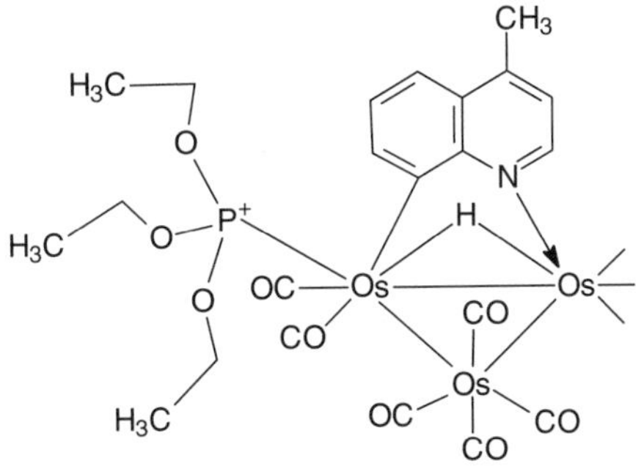

14

Metal-containing compounds are also being used as selective toxins. Brocard and coworkers have been studying ways to combat the malaria parasite, *Plasmodium falciparum*.[52] While there are drugs such as chloroquine that are able to inhibit the malaria parasite, resistance to these drugs is increasing. Since these parasites need blood, a strategy was worked out that combined the poison, chloroquine, with a bait, ferrocene that contains the iron necessary to produce hemoglobin. The compound called *ferroquine*, **15**, shown below, is much more potent than chloroquine at inhibiting the parasite in mice. It is active against both chloroquine-sensitive and chloroquine-resistant strains of *Plasmodium*. It is also safe and nonmutagenic.

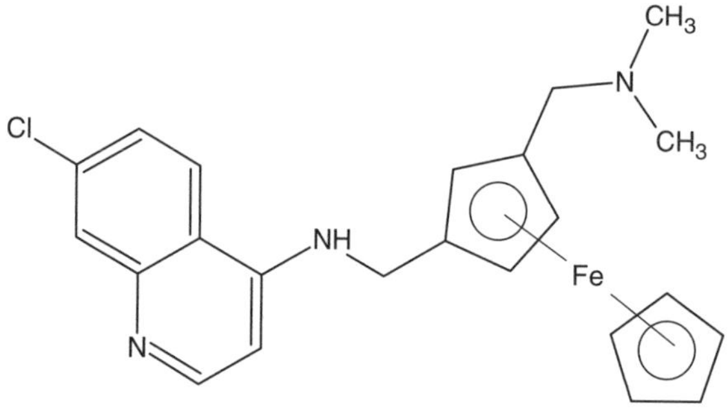

15

Metal-containing compounds are also used in other biomedical applications, independent of any biological response. Technetium-99m sestamibi, a hexakis(alkyl isocyanide) complex of ^{99}Tc, is one of the most important myocardial imaging agents. It is sold under its tradename of Cardiolite. Its structure, **16**, is given below, where only two of the six connecting ligands are given.

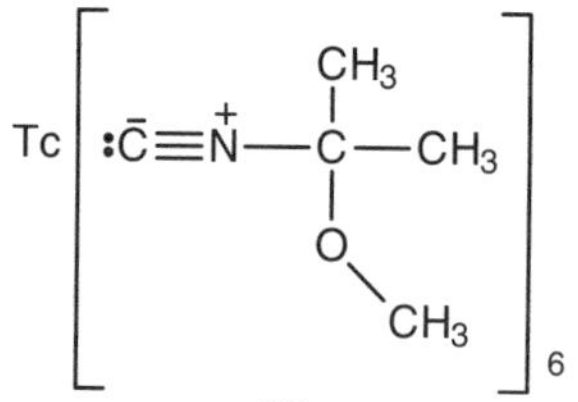

16

Half-sandwich metal-containing compounds are being studied for use in the radiopharmaceutical industry.[53,54] They are made from cyclopentadiene attached to a targeting biomolecule such as a tumor-specific peptide or a small molecule that binds to a central nervous system receptor. One of these, **17**, is depicted below. The half-sandwich complex is robust and lipophilic. Hopefully it will ferry the radioactive label across the blood-brain barrier. Again, emphasizing Tc, Alberto and coworkers have made a number of Tc compounds as possible replacements for Cardiolite. They have also been looking at Tc-containing materials as possible anticancer agents. One such representative structure is **17**. The use of ^{99}Tc in diagnostic medicine takes advantage of its gamma-ray emission. In about 20% of the decompositions, the metastable nucleus decays by internal conversion leading to the ejection of inner and outer-shell electrons that may be used for cancer therapeutic purposes if the radionuclide is near the DNA strand. Current efforts are aimed at looking at binding such to compounds with the DNA bases.

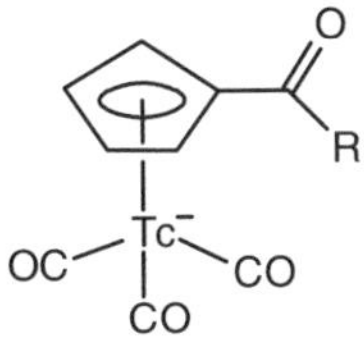

17

Metal-containing compounds are also being investigated as analytical tools in the biomedical arena. For instance, there is a need to develop quick ways to test for lithium. Lithium carbonate is an important drug in the treatment of bipolar disorder. The therapeutically useful range of lithium concentration in the blood is narrow, so monitoring is important. It is currently done using atomic emission spectroscopy, a time-intensive and costly technique. There are many researchers looking at this problem, and some of them are using metal-containing compounds as the sensing agents. Severin and coworkers are looking at a variety of Ru-, Rh-, and Ir-containing macrocyclic lithium receptors.[55] One such cage structure is depicted below. The oxygen-rich inner core tightly bonds the lithium ion. The group is able to selectively bind lithium ion from aqueous solutions even in the presence of many other metal ions, including solutions saturated with the sodium ion.

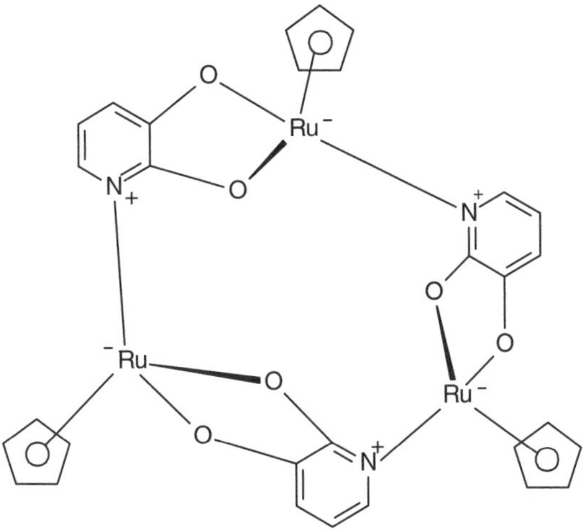

18

Along with their applications as analytical tools and for complexation within caged arrangements, metals are also being incorporated as the essential sites within a number of caged compounds for biomedical uses. Here, we will briefly describe one of these opportunities.

While MRI (magnetic resonance imaging) provides good imaging resolution, it often has a limited sensitivity. To increase the sensitivity, novel materials that have stronger proton relaxivity and higher MR signal enhancement at low concentration are being studied. Gadolinium chelate compounds are among the most widely employed magnetic resonance imaging, MRI, contrast agents.[56–59] They have been used to assist the assessment of abnormalities such as brain tumors and hepatic carcinoma.

Water-soluble multihydroxyl lanthanoid endohedral metallofulleronols have been synthesized and characterized for use as MRI contrast agents.[60] Endohedral metallofullerenes have been studied since the early 1990s.[61–63] One of the more important and novel electronic properties of these metallofullerenes is the so-called intrafullerene electron transfer from the encaged metal atoms to fullerene cages.[61–63] This allows their use as MRI contrast agents. The most widely studied of these compounds are the Gd endohedral metallofulleronols.[60] Preliminary studies are promising.[60] These materials have high longitudinal and transverse relaxivities for water protons, significantly higher than for corresponding lanthanoid chelate complexes. The large longitudinal relaxivity value is ascribed to the dipole–dipole relaxation together with a substantial decrease in the overall molecular rotational motion. The large transverse relaxivity is attributed to the so-called Curie spin relaxation.

V. SUMMARY

In summary, this chapter intends to convince the reader that metal-containing compounds, both small and large molecules, form the basis for wide-ranging, selective, and exciting contributions to biomedicine. Scientists might well view activities achieved with small-molecule metal-containing compounds as models for analogous polymer-containing materials that may have the advantages of being polymeric. These advantages might include restrictive movement, controlled release, modified solubilities, better specificity, greater stability, wider design of molecular capabilities, slower excretion rates, and multiple attachments. In the chapters that follow in this volume these potential advantages and opportunities are expressed in greater detail.

VI. REFERENCES

1. L. Friberg, G. Nordberg, in *Handbook on the Toxicology of Metals*, 2nd ed., Vol. I, *General Aspects*, G. Nordberg, V. Vouk, eds., Elsevier, Amsterdam, 1986.

2. B. Holmstedt, G. Liljestrand, *Readings in Pharmacology*, Raven Press, New York, 1981, pp. 22–30.

3. K. Hanusch, H. Grossman, K. Herbst, G. Rose, "Arsenic and Arsenic Compounds," in *Ullmann's Encyclopedia of Industrial Chemistry*, 5th ed., Vol. A3, Weinheim, Wiley, New York, 1985, pp. 113–115.

4. A. Albert, *Xenobiosis: Food, Drugs, and Poisons in the Human Body*, Chapman & Hall, New York, 1987, p. 1.

5. Mercury: A fact sheet for health professionals; www.orcbs.msu.edu/AWARE/pamphlets/hazwaste/mercuryfacts.html ref. To criminals in cinnabar mines.

6. B. Evers, S. Hawkins, G. Schulz, "Lead and Leading Poisoning in Antiquity," in *Ullmann's Encyclopedia of Industrial Chemistry*, Vol. A15, Weinheim, 1990, Wiley, New York, pp. 194–233.

7. J. Nriagu, *Lead and Leading Poisoning in Antiquity*, Wiley, New York, 1983.

8. M. Brannvall, R. Bindler, I. Renberg, O. Emteryd, J. Bartnicki, K. Billstrom, *Environ. Sci. Technol.* **33**, 4391 (1999).

9. See *J. Paint Technol.* **47**, 1088 (1975). The entire issue covered a special marine coatings symposium with seven major articles.

10. J. J. Zuckerman, ed., *Organotin Compounds: New Chemistry and Applications*, American Chemical Society, Washington, DC, 1976.

11. M. Cooksley, D. Parham, *Surf. Coat.* **2**, 280 (1966).

12. V. Castelli, W. Yeager, in *Controlled Release Polymeric Formulations*, D. R. Paul, F. W. Harris, eds., American Chemical Society, Washington, DC, 1976, p. 239.

13. See W. Aldridge in Ref. 10, p. 186, and M. Selwyn, in Ref. 10, p. 204.

14. A. T. Phillip, *Prog. Org. Coat.* **2**, 159 (1973).

15. T. Kealy, P. Pauson, *Nature* **168**, 1039 (1951).

16. G. Wilkinson, M. Rosenblum, H. Whiting, R. B. Woodward, *J. Am. Chem. Soc.* **74**, 2125 (1952).

17. R. B. Woodward, M. Rosenblum, H. Whiting, *J. Am. Chem. Soc.* **74**, 3458 (1952).

18. W. Troll, G. Witz, B. Goldstein, D. Stone, T. Sugimura, in *Carcinogenesis*, Vol. 7, E. Hecker et al., eds., Raven Press, New York, 1982, 593.

19. P. J. O'Brien, *Env. Health Persp.* **64**, 219 (1983).

20. T. Kensler, D. Bush, W. Kozumbo, *Science* **221**, 75 (1983).

21. I. Fridovich, *Annu. Rev. Biochem.* **44**, 147 (1976).

22. L. Oberey, G. Buettner, *Cancer Res.* **39**, 1141 (1979).

23. M. Sevilla, P. Neta, L. Marnett, *Biochem. Biophys. Res. Commun.* **115**, 800 (1983).

24. T. Kensler, M. Trush, *Biochem. Pharmacol.* **32**, 3485 (1983).

25. K. Horn, J. R. Dunn, *FEBS Lett.* **139**, 65 (1982).

26. E. W. Neuse, C. Mbonyana, in *Inorganic and Metal-Containing Polymeric Materials*, J. Sheats, C. Carraher, C. Pittman, M. Zeldin, B. Currell, eds., Plenum, New York, 1990, pp. 139–150.

27. P. Kopf-Maier, H. Kopf, E. W. Neuse, *J. Cancer Res. Clin. Oncol.* **108**, 336 (1984).

28. M. Wenzel, Y. Wu, E. Liss, E. W. Neuse, *Z. Naturforsch.* **43c**, 963 (1988).

29. C. C. Hinckley, P. Ostenburg, W. Roth, *Polyhedron* **1**, 335 (1982).

30. C. C. Hinckley, J. BeMiller, L. Strack, L. Russell, *Platinum, Gold, and Other Metal Chemotherapeutic Agents: Chemistry and Biochemistry*, S. J. Lippard, ed., American Chemical Society Symposium Series, Vol. 209, 1983, Chapter 21.

31. M. Nissila, *Scan. J. Rheumatol.* (Suppl.)**29**, 1 (1979).

32. I. Boussina, R. Lagier, H. Ott, G. Fallet, *Scan. J. Rheumatol.* **5**, 53 (1976).

33. C. C. Hinckly, S. Sharif, L. Russell, in *Metal-Containing Polymeric Systems*, J. Sheats, C. Carraher, C. Pittman, eds., Plenum, New York, 1985, pp. 183–195.

34. L. G. Donaruma, S. Kitoh, J. Depinto, J. Edzwald, M. Maslyn, in *Biological Activities of Polymers*, C. Carraher, C. Gebelein, eds., ACS, Washington, DC, 1982.

35. A. Winston, in *Bioactive Polymeric Systems: An Overview*, C. Gebelein, C. Carraher, eds., Chapter 21. Plenum, New York, 1985.

36. A. Winston, D. Varprasad, J. Metterville, H. Rosenkrantz, *Polymeric Materials in Medication*, C. Gebelein, C. Carraher, eds., Plenum, New York, 1984.

37. A. Winston, D. Varaprasad, J. Metterville, H. Rosenjrantz, *Polym. Sci. Technol.* **32**, 191 (1985).

38. A. Winston, D. Varaprasad, J. Metterville, H. Rosenkrantz, *J. Pharmacol. Exp. Ther.* **232**, 644 (1985).

39. P. Desaraju, A. Winston, *J. Polym. Sci., Polym. Lett.* **23**, 73 (1985).

40. H. Lossen, *Justus Liebigs Ann. Chem.* **150**, 314 (1869).

41. D. Coffman, U.S. Patent 2,402,604 (1946).

42. K. Norihiro, Jpn. Patent, 57102895, 1982.

43. D. Siegmann-Louda, C. Carraher, F. Pflueger, D. Nagy, J. Ross, in *Functional Condensation Polymers*, C. Carraher, G. Swift, eds., Kluwer, New York, 2002.

44. D. Siegmann-Louda, C. Carraher, Q. Quinones, G. McBride, *Polym. Mater. Sci. Eng.* **88**, 390 (2003).

45. C. Carraher, M. Naas, D. Giron, D. Cerutis, *J. Macromol. Sci.-Chem.* **A19**, 1101 (1983)

46. S. Top, E. Kaloun, A. Vessieres, I. Laios, G. Leclercq, G. Jaouen, *J. Organomet. Chem.* **350,** 643 (2002).

47. M. Mourits, E. De Vries, P. Willemse, K. Ten Hoor, H. Hollema, A. Van der Zee, *Obstet. Gynecol.* **97,** 855 (2001).

48. R. Witorsch, *Toxic Substance Mech.* **19,** 53 (2000).

49. T. Siden, A. Vessieres, C. Cabestaing, I. Laios, G. Leclercq, C. Provot, G. Jerard, *J. Organomet. Chem.* **500**, 637 (2001).

50. S. Top, H. El Hafa, A. Vessieres, M. Huche, J. Vaissermann, G. Jaouen, *Chem. Eur. J.* **8**, 5241 (2002).

51. E. Arica, D. Kolwaite, E. Rosenberg, K. Hardcastle, J. Ciurash, R. Duque, R. Gobetto, L. Milone, D. Osella, M. Botta, W. Dastru, A. Viale, I. Fieldler, *Organometallics* **17**, 415 (1998).

52. L. Delhaes, C. Biot, L. Berry, P. Delcourt, L. Maciejewski, D. Camus, J. Brocard, D. Dive, *ChemBioChem* **3**, 418 (2002).

53. J. Benard, K. Ortner, B. Springer, H.-J. Pietzsch, R. Alberto, *Inorg. Chem.* **42**, 1014 (2003).

54. F. Zobi, B. Springer, T. Fox, R. Alberto, *Inorg. Chem.* (in press).

55. M.-L. Lehaire, R. Scopelliti, K. Severin, *Inorg. Chem.* **41**, 5466 (2002).

56. H. Weinmann, R. Branch, W. Press, G. Wesbey, *Am. J. Radiol.* **142**, 619 (1984).

57. R. Brasch, H. Weinmann, G. Wesbey, *Am. J. Radiol.* **142**, 625 (1984).

58. P. Caravan, J. Ellison, T. McMurry, R. Lauffer, *Chem. Rev.* **99**, 2293 (1999).

59. D. Mitchell, *J. Mag. Res. Imag.* **7**, 1 (1997).

60. H. Kato, Y. Kanazawa, M. Okumura, A. Taninaka, T. Yokawa, H. Shinohara, *J. Am. Chem. Soc.* **125**, 4391 (2003).

61. R. Johnson, M. de Vries, J. Salem, D. Bethune, C. Yannori, *Nature* **355**, 239 (1992).

62. H. Shinohara, *Rep. Prog. Phys.* **63,** 843 (2000).

63. H. Shinohara, H. Saito, Y. Ando, T. Kodama, T. Shida, T. Kato, Y. Saito, *Nature* **357,** 52 (1992).

CHAPTER 2

Metal-Labeled DNA on Surfaces

Heinz-Bernhard Kraatz, Yitao Long, and Todd C. Sutherland

Department of Chemistry, University of Saskatchewan, Saskatoon, Saskatchewan, Canada

CONTENTS

I.	INTRODUCTION	20
II.	FERROCENE NUCLEOTIDES	20
III.	FERROCENE-DNA CONJUGATES	22
IV.	OTHER METAL-DNA CONJUGATES	34
V.	METALLATED DNA	36
	A. Cu-DNA	36
	B. M-DNA	37
VI.	SUMMARY	43
VII.	ACKNOWLEDGMENT	43
VIII.	REFERENCES	43

Macromolecules Containing Metal and Metal-Like Elements,
Volume 3: Biomedical Applications, edited by Alaa S. Abd-El-Aziz,
Charles E. Carraher Jr., Charles U. Pittman Jr., John E. Sheats, and Martel Zeldin
ISBN: 0-471-66737-4   Copyright © 2004 John Wiley & Sons, Inc.

I. INTRODUCTION

The electron transfer (ET) properties of biomolecules have attracted significant attention, largely driven by the desire to develop devices able to detect mutations or to prepare nanoscale biological devices, such as molecular wires.[1] In particular, ET studies of double-stranded DNA (ds-DNA) has become a popular area of research. Although results point toward hopping and G-to-G hole transfer and or more complex combination mechanisms,[2] the mechanistic details remain largely elusive.[3] In addition, the results were often confusing and contradictory. Some reports claimed that ds-DNA is a conductor and mediates ET via the π-stacked basepairs while other reports suggested that DNA is an insulator under certain conditions.[4] In this chapter, we are critically reviewing the current "state of the art" with particular focus on the use of self-assembled monolayers (SAMs) of DNA. Evaluation of the ET kinetics and exploitation of the DNA's inherent ability to recognize other DNA sequences both specifically and reversibly for applications as biosensors is explored. In this regard, metal-DNA conjugates and adducts of various flavors are useful,[5] and offer the potential for the electrochemical detection of DNA hybridization events and DNA mismatches without the need to label the analyte DNA.

II. FERROCENE NUCLEOTIDES

For purposes of simplicity, we will differentiate between conjugates of ferrocene (Fc) with DNA and other metal-containing conjugates.

For the site-specific covalent labeling of DNA oligonucleotides with Fc derivatives, two synthetic ways can be distinguished: (1) the direct coupling of Fc derivatives, for example, an active ester of Fc carboxylic acid, to an end-tagged primary amine on the DNA oligonucleotide in solution and (2) the synthesis of a nucleoside phosphoamidite derivative containing the Fc redox probe and incorporating it at any site during the standard automated DNA solid-phase synthesis. Fc carboxylic acid is readily converted into the active esters of hydroxybenzotriazole or the succinimide active ester in the presence of HOBt or HOSu, respectively (Scheme 1).[6]

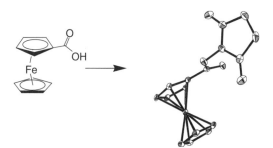

Scheme 1 Synthesis of ferrocene succinimide active ester from ferrocene carboxylic acid and succinimide in the presence of HOSu and a carbodiimide.

The active esters are stable and can be conveniently stored and used as stoichiometric Fc delivery reagents. Houlton and coworkers[7] reported the synthesis of ferrocenylthymidine derivatives via the Pd-catalyzed cross-coupling between ethynylferrocene or vinylferrocene and 5-iodo-2'-deoxyuridine. The synthesis is shown in Scheme 2.

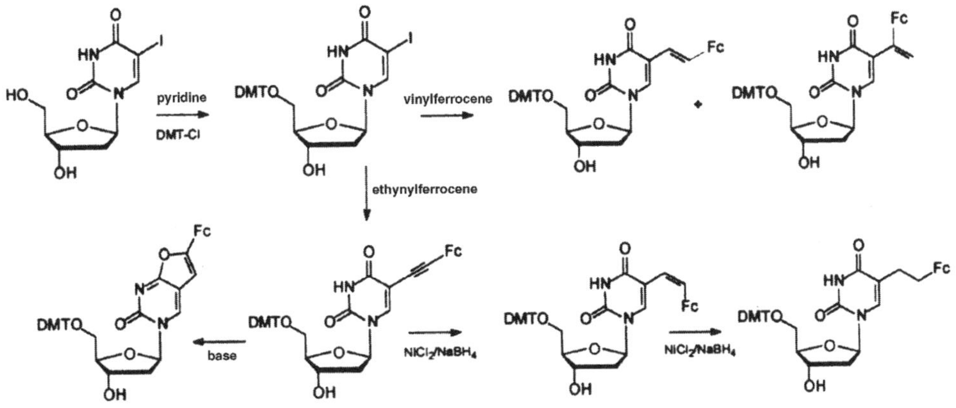

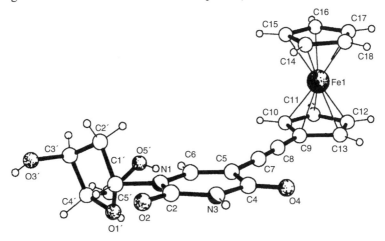

Scheme 2 Synthetic route to C5-substituted Fc thymidine derivatives (taken with permission from Ref. 7).

Figure 1 shows the crystal structure of 5'-dimethoxytrityl-5-ethynylferrocene-2'-dexyuridine. The Fc group is essentially coplanar with the base, allowing for facile electronic communication between the Fc group and the base. Importantly, this compound was incorporated site-specifically into DNA oligonucleotides by solid-phase methods, thus giving redox-labeled oligonucleotides. The Fc-labeled 20-mer oligonucleotide exhibits a $E°$ of 286 mV (by DPV, at a scan rate of 20 mV/s, with a Ag^+-quasi-reference electrode, tungsten counter-electrode, and a gold disk working electrode in 0.1 M TEAA buffer at pH 6.5).

Figure 1 The structure of 5'-dimethoxytrityl-5-ethynylferrocene-2'-dexyuridine (taken with permission from Ref. 7).

Other groups have made use of this strategy and were able to introduce the Fc-labeled nucleotides into an oligonucleotide either by solid state reactions or by the polymerase chain reaction (PCR).

III. FERROCENE-DNA CONJUGATES

A number of Fc-DNA conjugates were reported in the literature, and a selected number of the reports are summarized in Table 1. Fc-labeled DNA constructs have emerged as an effective tool to investigate DNA hybridization and offer a convenient method for DNA mismatch detection by differential hybridization.

Scheme 3 shows a typical procedure using the solid-state phosphoramidite protocol used to synthesize a DNA oligonucleotide of a specific sequence on a glass bead. As the final step of the solid-phase synthesis, an amino linker is attached to the oligonucleotide. After cleavage of the construct from the glass bead, the Fc redox group is connected to the 5′-amino-tagged oligonucleotide via an amide bond. Other groups have attached the Fc redox group via a phosphoester linkage. The conjugate is stable and is commonly purified by HPLC. In this particular case, the conjugate synthesized possess a disulfide linkage, which is used to attach the oligonucleotide to a gold surface.

Table 1 List of Fc-DNA Conjugates

Entry	Compound	Remark	Ref.
1	5′-Fc- and 3′-thiol modified nucleotide		5b
2	5′-Fc nucleotide	22-55 pmol/cm^2 +645 mV vs. Ag/AgCl	8
3	Fc-tethered dUTP	Biocatalytic amplification of DNA detection	10
4	3′-Fc-labeled oligonucleotide	Flexibility of DNA on Au	5a
5	substituted Fc on nucleotides	Sinoidal voltammetry	20
6	Fc-modified nucleotides	Molecular diagnostics platform	14
7	Fc-dUTP	Enzymatic redox labeling of DNA	12
8	Fc-oligonucleotides	DNA detection	9
9	3′(5′)Fc-oligonucleotide	Comparision ET rates	16
10	Fc-DNA stem loop	DNA detection 10 pM detection limit	13

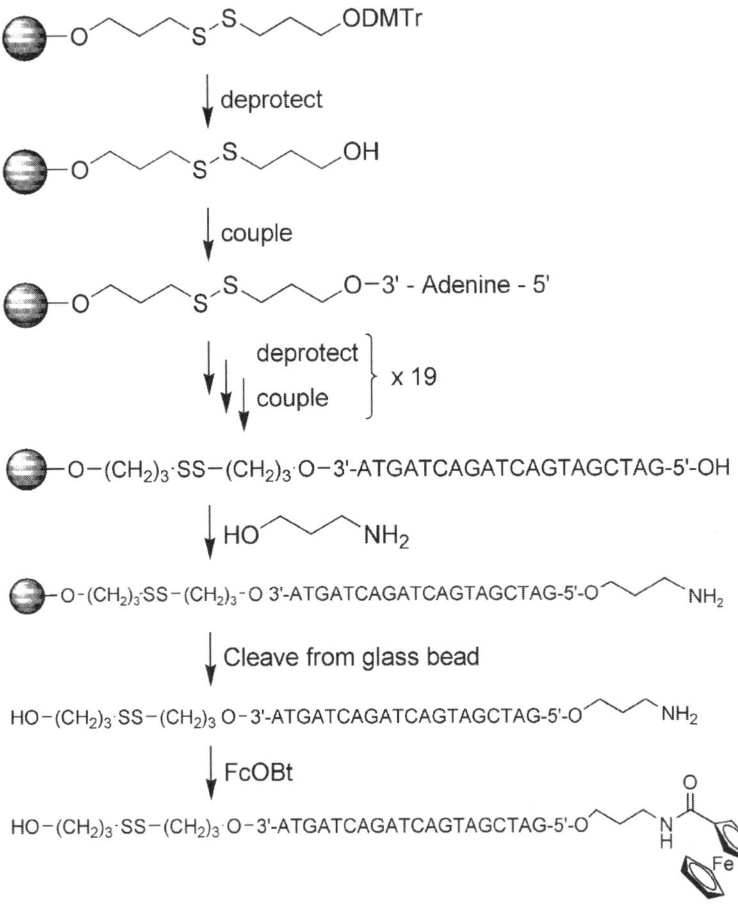

Scheme 3 Solid-state synthesis of a 5'-amino-tagged oligonucleotide followed by 5'labeling of the oligonucleotide with the –OBt active ester of ferrocene carboxylic acid.

As mentioned above, the redox label can be introduced as a substituent on a nonnatural nucleoside and then incorporated using solid state synthetic procedures.

Driven by the potential application of DNA in electrochemical biosensing and as a self-assembling molecular wire, SAMs on conducting substrates such as gold have been studied by a number of groups. Mirkin, Letsinger, and coworkers reported a nucleotide labeled via a 5'-ferrocene phosphoester and a 3'-thiol for chemisorption onto gold surface.[5b] The system is redox-active, allowing the characterization of the modified surface by electrochemical techniques. The pendant Fc group exhibits a fully reversible one-electron redox wave with a formal potential $E^{o'}$ of 220 mV (vs. Ag/AgCl) (see Fig. 2).

The Fc group provides a convenient method for determining the surface concentration of the adsorbate on the gold surface (3.0×10^{-10} mol/cm^2). The resulting footprint per molecule is given as 9.8 Å, which is larger than for Fc-alkylthiols on gold. Single-stranded DNA on gold surfaces is very flexible but becomes significantly more rigid by hybridization of the complementary strand.[5a]

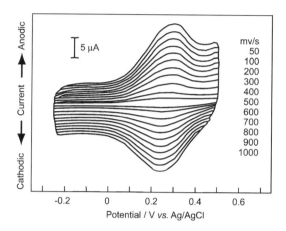

Figure 2 Cyclic voltammogram showing the electrochemical behavior of Mirkin and Letsinger's Fc-thymidine-3′-thiol conjugate on a gold electrode. The peak currents scale linearly with scan rate as expected for a molecule chemisorbed onto the surface. (Reproduced with permission of The Royal Society of Chemistry from Ref. 5b.)

Ihara et al.[8] presented a gene-sensing system using a Fc-modified single-strand oligonucleotide and a single-strand oligonucleotide possessing five successive thiophosphate groups. These are readily incorporated by conventional solid-phase DNA synthesis. The thiophosphate groups anchor the molecule onto a gold electrode. The surface-supported strand carries a sequence that is complementary to a target strand.

The target DNA strand can then be hybridized to the surface-supported strand. This is followed by the hybridization of a Fc-oligonucleotide to a complementary sequence on the ss-DNA target strand. The hybridization event forming the ternary complex will result in a detectable electrochemical signal (Fig. 3).

A significant anodic peak due to oxidation of the Fc moiety was observed for the electrode treated with the target DNA. The peak potential (E_p) of 645 mV (vs. Ag/AgCl) for the monolayer was exceedingly positive compared with the redox potential of ~430 mV in bulk solution. Importantly, the monolayer is stable even to heating to 80°C without loss of the electrochemical signal.[7] The electrochemical results compare well to those of Letsinger and Mirkin, who also observed a shift to more positive potential on surface binding.[5b] Upon raising the temperature above the melting point of the double helix, dehybridization will occur, resulting in a loss of the Fc target signal. A background current, or residual current, often observed in these studies has been attributed to the non-specific binding of single-stranded DNA to the electrode surface. The proposed method is an intriguingly simple and selective way to detect DNA hybridization. Using a 19-mer as the target strand and a Fc-labeled 12-mer as a probe strand, it was possible to detect mismatches by DPV. Mismatches will often have significantly lower melting points. Dehybridization from the surface-bound strand will occur at temperatures at which a fully matched strand remains hybridized to the surface-bound strand. Using this strategy, Ihara and coworkers were able to detect single-nucleotide mismatches.[9] However, the background currents appear high in the reported data.

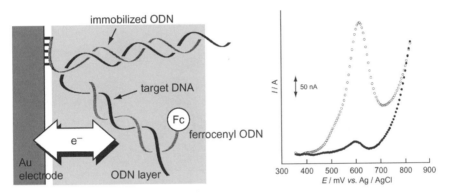

Figure 3 Left: schematic representation of the interaction of the Fc-labeled target strand with the thiophosphate labeled oligonucleotide on a gold surface. Right: DVP response of the Fc-labeled target strand hybridized onto the surface strand (○) compared to the background signal (•) due to the nonspecific interaction of the non-complementary Fc-oligonucleotides with the electrode surface. (Reproduced with permission of The Royal Society of Chemistry from Ref. 8.)

Willner and coworkers[10] reported an important extension of this work by having a bioelectrocatalytic amplification of the DNA detection process.

Figure 4 shows the amplified electrochemical detection of viral DNA. Viral DNA interacts with a thiolated oligonucleotide chemisorbed on a gold surface. This oligonucleotide is complementary to viral DNA (M13Φ), and thus the viral DNA will bind to the DNA film. The partially double-stranded DNA surface was then allowed to interact with a nucleotide mixture that included Fc-labeled UTP in the presence of Klenow fragment I polymerase. Under the conditions chosen, the polymerase replicated the viral DNA incorporating the Fc-labeled DNA at efficiencies of about 59% (as judged by quartz crystal microbalance measurements). In total it was estimated that each replica contained about 350 Fc-labeled bases. In this system, the Fc units acted as redox mediators between the electrode surface and glucose oxidase, which in turn catalyzed the conversion of glucose to gluconic acid. Figure 5a shows the DPV of the Fc-labeled replica as it is being incorporated into the viral DNA replica by the polymerase. After 60 min, the DPV showed a redox signal characteristic of the Fc group. The current scales appropriately and reaches a saturation as all replicas on the surface are complete. Figure 5b shows the electrocatalytic current of the replica surface with glucose oxidase in solution in the absence (point *a*) and presence of glucose.

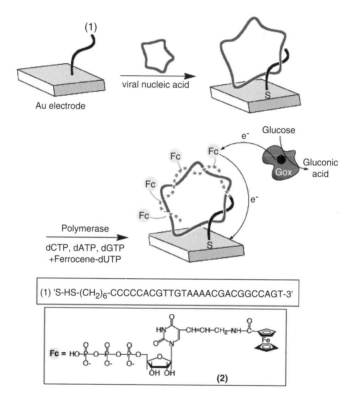

(1) 'S-HS-(CH₂)₆-CCCCCACGTTGTAAAACGACGGCCAGT-3'

Figure 4 Fc-labeled UTP can be incorporated into a copy of viral DNA by Klenow I polymerase using a viral DNA hydridized to a complementary strand of surface-bound ss-DNA. The Fc-labeled copy can then act as a redox relay for the biocatalytic oxidation of glucose to gluconic acid by glucose oxidase. (Reprinted with permission from Ref. 10. Copyright 2002 American Chemical Society.)

This work demonstrates the ability to use redox-active DNA replicas to activate bioelectrocatalytic cascade reactions and makes use of DNA recognition events to bind the DNA to a transducer-bound DNA target. Thus, this work represents an interesting new facet in DNA bioelectronics. Of particular interest is the Fc-nucleotide used in Willner's work. It involves the use of a Fc active ester (Fc-succinimide), which can then be coupled to propargylamine to form the corresponding amide, followed by coupling to uridine. Beilstein and Grinstaff have reported the convenient used of Fc-propargylamide to label a column bound 5-iodo-2′deoxyuridine during standard oligonucleotide synthesis.[11] Importantly, the melting point of ds-DNA is not significantly influenced by the presence of the Fc group. Simultaneous with Willner's report, King and coworkers[12] reported the full synthesis of two ferrocene-labeled bases 5-(3-ferrocenecarboxamidopropenyl-1) 2′-deoxyuridine 5′-triphosphate (Fc1-dUTP) and 5-(3-ferroceneacetamidopropenyl-1) 2′-deoxyuridine 5′-triphosphate (Fc2-dUTP), which were incorporated into a DNA sequence by polymerases (Klenow and T4 DNA polymerases). Both redox labels display a fully reversible redox behavior in solution with $E_{1/2}$ of 398 mV for Fc1-dUTP and 260 mV for the ferroceneacetic acid derivative Fc2-dUTP (vs. Ag/AgCl).

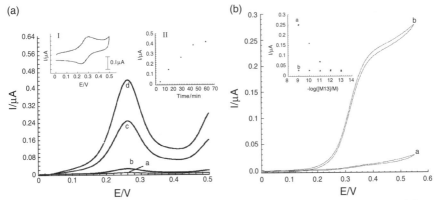

Figure 5: (a) Differential pulse voltammogram (DPV) showing the increase in Faradaic current as the Fc-labeled replica are being synthesized by the polymerase; (b) cyclic voltammograms (CVs) of Fc-labeled DNA in the presence of glucose oxidase. Point *a* shows the CV in the absence of the substrate glucose; and Point *b* shows the CV in the presence of glucose, resulting in a large catalytic current. (Reprinted with permission from Ref. 10. Copyright 2002 American Chemical Society.)

Importantly, King succeeded in the formation of a Fc-labeled 998-bp (base-pair) construct by PCR using T4 DNA polymerase in the presence of Fc1-dUTP. Incorporation of the redox label shows that Fc1-dUTP is suitable as a substrate for PCR. In contrast, Fc2-dUTP acts predominantly as a terminator in the PCR. The melting behavior of a 37-mer duplex containing five Fc1-dU residues reveals that the labeled nucleotide induces only a modest helix destabilisation, with $T_m = 71°C$ for a labeled duplex versus 75°C for the corresponding nonlabeled ds-DNA construct. King reports that the Fc-labeled DNA is detected at femtomolar levels by HPLC using a coulometric detector. Thus, it must be emphasized that the incorporation of the Fc label by PCR and its facile and cost-effective electrochemical detection should promote the use of this technique in nucleic acid analysis and may replace the more costly fluorescence-based detection systems.

Heeger and coworkers[13a] reported a 5'-ferrocene-labeled single-strand DNA construct with a 3'-sulfhydryl group for attachment to a gold surface (Fig. 6). The DNA carries a self-complementary sequence that is able to form a stem loop structure. In this conformation, the Fc group is close to the electrode surface. On presentation of a complementary DNA sequence in solution, the stem loop will open and the Fc-labeled target strand will hybridize with the complementary DNA sequence. This significant conformational change of the ds-DNA significantly increases the separation between the Fc group and the electrode surface, which will change the electron transfer properties of the assembly.

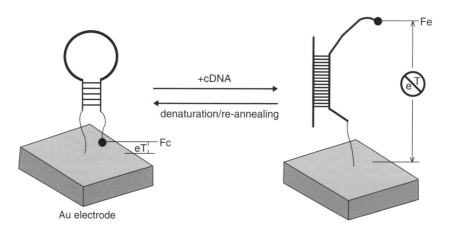

Figure 6 Heeger's stem loop for the detection of DNA hybridization.[13]

In this report, Heeger and coworkers claim that the biosensor system allows the measurement of the hybridization efficiency at concentrations down to 10 pM. The importance of this method comes from the extremely low detection limit as well as the "reagentless" detection. Two years before Heeger, a similar design was proposed and evaluated by Xi Hu, a Ph.D. student in Grinstaff's group at Duke University.[13b] Their design is analogous to that reported by Heeger working with a hairpin design in which coordination of the complementary strand will change the distance between the Fc electrophore and the electrode surface. Hu's design anchors the Fc-labeled DNA hairpin on a surface via standard thiolate chemistry. The construct is present in a thioundecanoic acid diluent. Hu reports that complementary and non-complementary strand can be distinguished on the basis of the Fc oxidation potential, the rate of electron transfer, and the changes in current.

Umek et al.[14] reported the application of AC voltammetry for the electrochemical detection of oligonucleotides on printed gold microarrays and outlined a highly selective detection system for applications in molecular diagnostics. The format has a printed gold electrode with 14 separate exposed gold electrodes, which are coated with a SAM containing $3'$-thiol terminated DNA capture probes. The electrode diameter is either 250 or 500 µm. Unlabeled nucleic acid targets are immobilized on the surface of the SAM through sequence-specific hybridization with the DNA capture probe. A separate "signaling probe," containing Fc-labeled nucleotides, which is complementary to the target in the region adjoining the capture probe is held in close proximity to the SAM in a sandwich complex (Fig. 7). The Fc label is introduced into a series of signaling probes via a modified Fc-ribose-labeled adenine base. The sequences of the signaling and capture probes are designed to be complementary to adjacent regions of the target.

The SAM allows electron transfer between the immobilized Fc on the signaling probe and the gold surface, while at the same time insulating the electrode from soluble unbound signaling probes, which may otherwise cause a significant background signal. The researchers succeeded in achieving single basepair mismatch discrimination based on differential hybridization, in an approach outlined above. However, it should be pointed out that melting is not necessary for mismatch detection, as the system shows a significant distance dependence for signal detection. The system presents a significant step forward toward a reliable electrochemical biosensor platform that may be suitable for disease diagnostics and pharmacogenetics. In a set of real-life experiments, single-nucleotide polymorphism discrimination is demonstrated using a genotyping chip for the C282Y single-nucleotide polymorphism associated with hereditary hemochromatosis. Furthermore, the technique shows utility for infectious disease diagnostics, such as HIV diagnostics.

The electrochemical detection of mismatches and DNA hybridization relies on ET from the redox reporter to the transducer surface. Thus, gaining a thorough understanding of the ET kinetics involved in the sensing process is crucial in order to optimize the detection, and eliminate potential flaws of this method.

Based on earlier results by Barton and coworkers,[15] who reported that the electron transfer kinetics will exhibit a directional dependence, Kraatz and coworkers[16] reported a comparative study of the ET through two configurations of ds-DNA. For this purpose, two Fc-labeled DNA conjugates and their complementary strand were prepared by solid phase synthesis:

1: $5'$-Fc-NH-$(CH_2)_3$-AACTACTGGGCCATCGTGAC-$3'$-$(CH_2)_3$-S-S-$(CH_2)_3$-OH

2: $3'$-TTGATGACCCGGTAGCACTG-$5'$

3: $5'$-AACTACTGGGCCATCGTGAC-$3'$-$(CH_2)_3$-S-S-$(CH_2)_3$-OH

4: $3'$-Fc-NH-$(CH_2)_3$-$3'$-TTGATGACCCGGTAGCACTG-$5'$

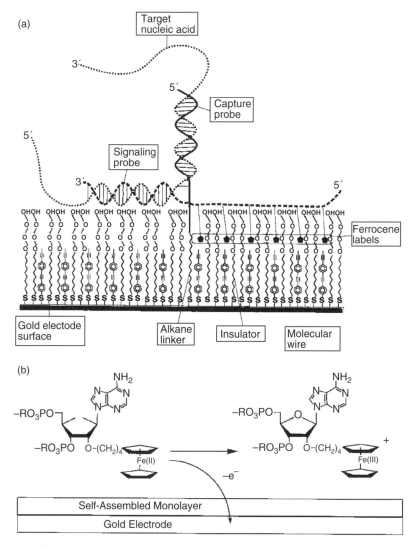

Figure 7 Umek's detection system using a SAM containing 3'-thiol terminated DNA capture probes and Fc-labeled signaling probes.[14]

The Fc label was introduced as an amide using Fc-OBt,[6] an active ester that readily reacts with available amino groups. One construct had the Fc group at the 5' terminus of the DNA, whereas the other had the Fc at the 3' terminus. The DNA constructs were purified by RP-HPLC and characterized by MALDI-TOF MS and UV–vis spectroscopy. Typical UV–vis spectra for Fc-labeled ds-DNA and Fc-labeled ss-DNA are shown in Figure 8. After hybridization of the Fc-labeled DNA strand to its complementary strand in 20 mM Tris buffer (pH 8.7) for 24 h at room temperature, the Fc-labeled ds-DNA was exposed to gold microelectrodes of 50 μm diameter (0.05 mM ds-DNA in 20 mM Tris buffer at pH 8.7) at room temperature for 5 days. XPS measurements of the surface clearly showed S_{2p} peaks at 162 and 163 eV, in a 2–1 ratio, which are indicative of Au-bound thiolates.[17] By monitoring the attenuation of the Au_{4f} peaks, the film thickness was calculated to 54(2) Å for both configurations,[18] which compared well with ellipsometry data on these films.

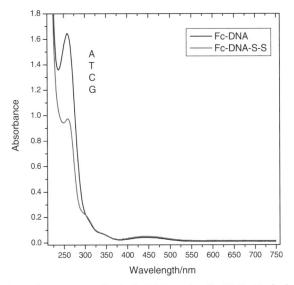

Figure 8 Typical absorption spectrum for ds-Fc-DNA and ss-Fc-DNA. Both show a ferrocene-based absorption maximum at $\lambda=440$ nm.

The results of the electrochemical studies are shown in Figure 9. Cyclic voltammetric studies of the **1–2** and the **3–4** configurations show that the peak current scales linearly with the scan rate, confirming a surface bound Fc-DNA. Furthermore, the CVs for both configurations exhibit a quasireversible redox behavior with the anodic to cathodic peak current ratios near unity. The CV of the **1–2** construct has a peak width at half-maximum of 102(10) mV, which is close to ideal redox behavior of 90 mV while that of the **3–4** construct is significantly larger (125(7) mV), indicating that the Fc groups in the **1–2** hybrid act as independent isolated redox centers in identical environments, whereas the CV results indicate the possibility of lateral interactions or microenvironments in the **3–4** hybrid.

The redox potential of Fc is known to be sensitive to its environment and as such has been employed in several sensor development schemes. On the basis of the monolayer structural evidence, the Fc environment in both films should be very similar and thus should result in similar formal potentials $E^{\circ'}$. The flexibility of the C_3 is expected to average small local differences. However, the electrochemical results show a difference in the redox potentials for the two configurations of 29(14) mV (**1–2** monolayer: $E^{\circ'}=408(8)$ mV; **3–4** monolayer: $E^{\circ'}=437(11)$ mV). This corresponds to a $\Delta\Delta G$ value of 2.8(1.4) kJ/mol between the two films. This demonstrates that the Fc groups in both films are in different microenvironments such that the Fc group is less accessible to electron transfer in the **3–4** film. The implications are that electron transport to the Fc group is more facile when the redox probe is positioned 5′ compared to 3′. The difference in $E^{\circ\prime}$ values could be attributed to geometry and orientation of the Fc due to 5′ or 3′ linkage leading to a different interaction with the basepair stack. In addition, the ΔE_p values increase when moving from a **1–2** film to a **3–4** film, indicating a difference in the ET kinetics and a lower k_{ET} for the **3–4** monolayer. ET rates of the two configurations were calculated from the CV of both films using the Butler–Volmer methodology.[19]

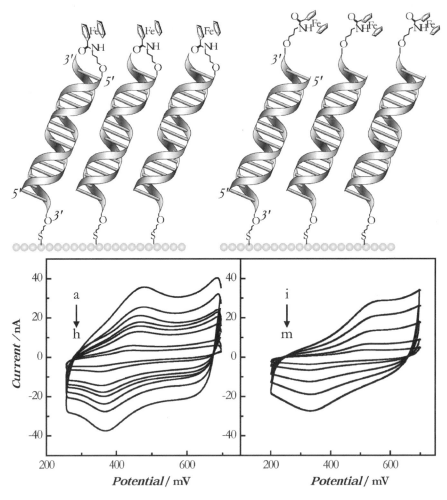

Figure 9 Representation of the conjugate **1–2** bound to a gold surface. In the **1–2** construct, the Fc label is on the same strand as the thiolate whereas in the **3–4** construct, the Fc is not on the surface-bound strand but on the complementary strand of the thiolate. CVs: (*a*) 17.5 V/s; (*b*) 12.5 V/s; (*c*) 10 V/s; (*d*) 8.5 V/s; (*e*) 7.5 V/s, (*f*) 6 V/s; (*g*) 2 V/s; (*h*) 1.36 V/s; (*i*) 15 V/s; (*j*) 12 V/s; (*k*) 8 V/s; (*l*) 4 V/s; (*m*) 2 V/s. Insets: linear relationship between scan rate and peak current.

The calculated ET rate constants for the **1–2** and **3–4** films are $115(15)$ s^{-1} and $25(9)$ s^{-1}, respectively. Therefore, not only are the thermodynamics in favor of electron transport if the Fc group is on the same strand as the thiolate connecting it to the surface, but also the kinetics favor **1–2** for ET. It is important to stress at this point that the 3′-thiolate linkage is the same for both configurations. A close examination of the results provides more clues toward solving the ongoing ET debate in DNA. Both configurations should give indistinguishable results if ET proceeds entirely in a through-space model. A report in 2002[15] claimed that ET is favored in the 5′ to 3′ direction (Scheme 4a); however, this is unlikely to be the case of the two DNA constructs **1–2** and **3–4**. In the electrochemical measurements, a directional preference would introduce a significant asymmetry in the peak shape for the forward or backward reaction. This is not the case, and thus a model proposing directional preference must be discarded.

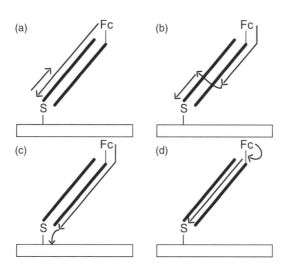

Scheme 4 Schematic representation of the **3–4** construct on a gold surface, showing potential ET pathways through the construct: (*a*) 5′-3′ strand preference, (*b*) barrier to interstrand ET, (*c*) barrier for ET through space to Au from terminal basepair; (*d*) specific Fc orientation effects.

Three additional models are shown in Scheme 4. Scheme 4b assumes no directional preference ET in the DNA constructs. Thus, the only additional barrier to electron transfer is ET between the two strands (interstrand jumping), which results in the **3–4** construct having a lower k_{ET} and symmetrical CVs. Scheme 4c assumes that interstrand jumping is forbidden and as such, the rate-limiting step occurs when the electron tunnels to and from the basepair proximal to the gold surface. Scheme 4d assumes that ET is not confined to one strand but rather, the rate-determining step is the feeding in of the electron into the first basepair. It is possible that there is orientation or geometric effects that predispose the 5′-labeled Fc to be more accessible to the basepairs than the 3'-labeled Fc. This is supported by the different $E^{\circ'}$ values observed, as discussed above. Of course, a combination of strand jumping and Fc orientation could be operating simultaneously. Only additional studies will shed more light onto the details of the ET process.

In contrast to the systems presented above, Kuhr and coworkers[20] use a Fc tag for DNA sequencing applications. Interestingly sinusoidal voltammetry is used to selectively identify four unique Fc-redox sequences that are covalently attached to the 5' end of a 20-base sequencing primer. The electron-donating or withdrawing character of the substituent alters the half-wave potential of the modified Fc group. The Fc-tagged DNA oligonucleotides are separated by sieving polymer capillary electrophoresis with an efficiency of 2×10^6 theoretical plates. The Fc groups are then detected electrochemically, allowing for "low-resolution" DNA sequencing.

Fang and coworkers[21] attached ferrocene carboxaldehyde to an amino-tagged oligonucleotide and also showed it to be redox-active. The label was attached to denatured calf thymus DNA, and it was shown that binding of the complementary strand can be detected down to 5×10^{-9} M.

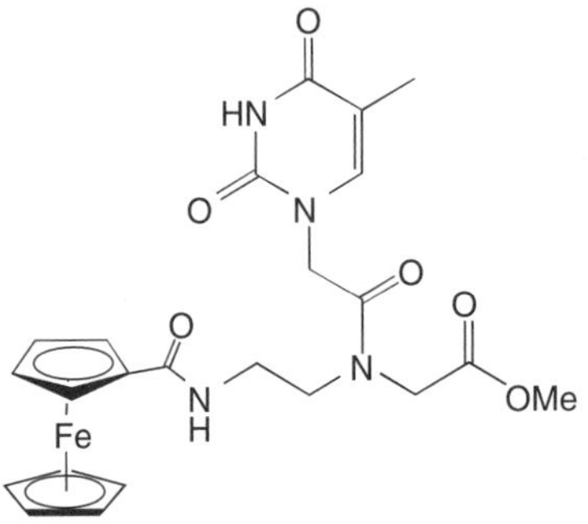

An interesting variation on the DNA theme was reported by Hess and Metzler-Nolte,[22] who reported the synthesis of Fc-peptide nucleic acids (Fc-PNA), which represent a family of DNA analogs with related molecular recognition properties. Other metal-containing redox probes were also attached to PNA monomers. The PNA consists of a pseudopeptidic backbone to which the nucleobases are attached though a methylene carbonyl spacer. PNA has a high affinity for ss and ds-DNA and readily forms double and triple helices in the process. Metzler-Nolte report the incorporation of a Fc tag to the N terminus of a PNA monomer, effectively bringing together peptide and DNA labeling strategies.[23]

IV. OTHER METAL-DNA CONJUGATES

In 1995 Meade and Kayyem[24] reported the synthesis of an electron donor–acceptor pair of 5'-ruthenium-modified nucleosides in which a $(bipy)_2Ru(II)$

or a $(H_3N)_4pyRu(II)$ is coordinated to an aminofuranose ring. The resulting labeled nucleosides are stable to HPLC purification and were subsequently incorporated into oligonucleoticles by solid-phase synthesis. More recent results by Meade and coworkers are related to the use of chelating nucleosides, such as 5-*O*-(4,4'-dimethoxytrityl)-2'-iminomethylpyridyl-2'-deoxyuridine, which readily binds $(acac)_2Ru(II)$ and $(bipy)_2Ru(II)$ fragments.[25] In contrast to the Ru(II)-labeled aminofuranose systems, the iminoethylpyridyl-labeled nucleosides appear to be electronically coupled, as judged from the absorption spectra of the complexes and their redox behavior. This is rationalized by the proximity of the metal center to the base. These Ru(II) nucleosides are readily incorporated into oligonucleotides by phosphoramidite chemistry, giving 5'-labeled DNA oligonucleotides.[26] Once incorporated into the oligonucleotides, the electronic structure appears unperturbed. Furthermore, the stability of the DNA duplex is not adversely influenced by the presence of the label. Thus, the labeled nucleosides are a valuable addition to the manifold of redox-labeled nucleosides and offer the potential to investigate the ET characteristics in DNA constructs in solution and on the surface.

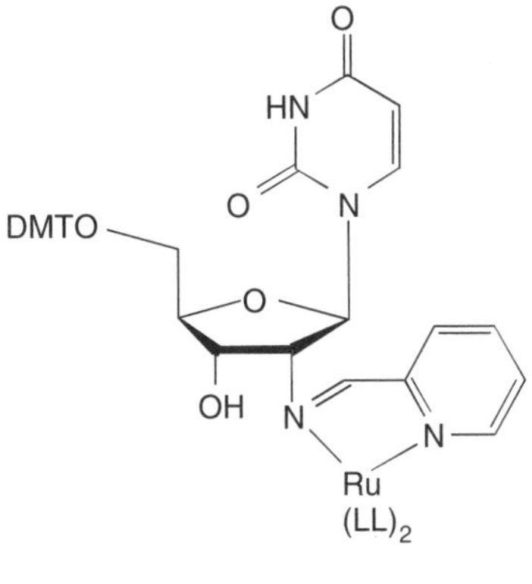

Related is the approach described by Grinstaff[27] using 5'-Ru(bipy)$_3$-labeled oligonucleotides. 4-methyl-2,2'-bipyridine-4'-carbonylethanolamide and Ru(bipy)$_2$Cl$_2$ readily react to give the corresponding Ru complex, which can then be converted into a complex suitable for phosphoramidite chemistry.

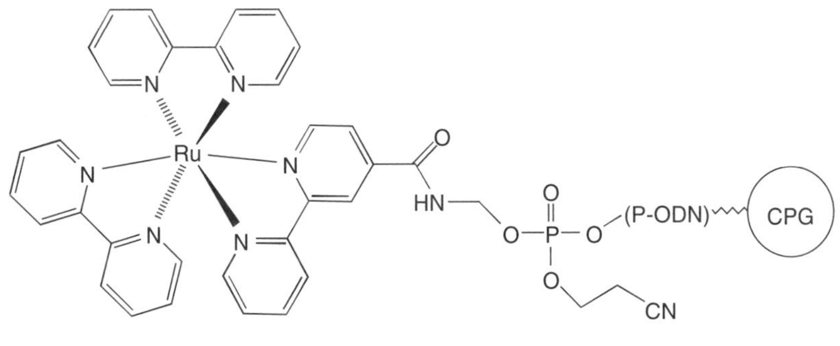

Starting from 5-iododeoxyuridine, Hurley and Tor[28] synthesized 1,10-phenanthroline-labeled nucleosides, containing a $[(bipy)_2M(3\text{-ethynyl-1,10-phenanthroline})]^{2+}$ (M=Ru, Os) moiety covalently attached to the 5 position of the 2'-deoxyuridine (Scheme 5). Both complexes exhibited a fully reversible electrochemistry with $E_{1/2}=0.993$ V for $Ru^{2+/3+}$ and 0.560 V for $Os^{2+/3+}$ (vs. Ag/Ag^+, in MeCN). In all cases, the nucleosides can be converted into phosphoamidites to allow for facile incorporation into oligonucleotides by automated solid-phase synthesis.

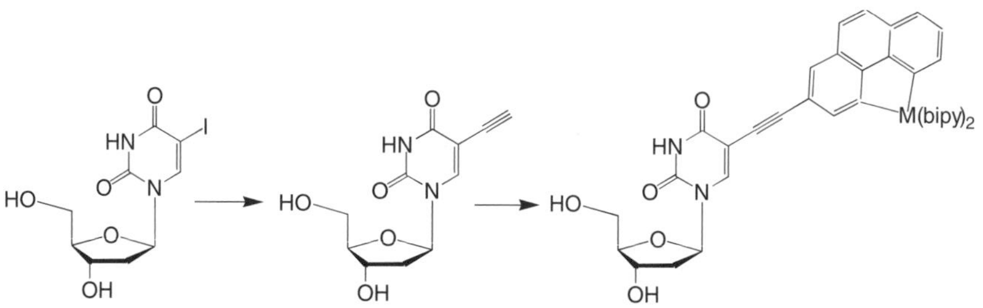

Scheme 5 Synthesis of metal-labeled phenanthroline nucleosides according to Hurley and Tor.[28]

Barton[29] reported the synthesis of the DNA intercalator Δ-$[Rh(phi)_2(bipy')]^{3+}$ tethered to a 20-mer oligonucleotide. In previous studies it was shown that photoinitiated ET between metallointercalators that are bound to DNA are fast and efficient compared to unbound systems in which the intercalator simply intercalates into the DNA base stack.[30] Thus, these systems should be ideally suited to investigate the ET properties of metal-modified DNA assemblies in solution.

A further development was reported by Kowalski and Zakrzewski,[31] who synthesized a series of η^5-CpFe(CO)$_2$-labeled (Fp-labeled) nucleotides with 2',3',5'-tri-O-acetyluridine, 3',5'-di-O-acetylthymidine, and 5'-O-(4,4'-dimethoxytrityl) thymidine.

V. METALLATED DNA

A. Cu-DNA

Shionoya and coworkers[32] reported a DNA-like material in which artificial hydroxypyridone nucleotides are employed to bind to copper ions. In this form, the Cu(II) ions replace the hydrogen-bonded basepairing in DNA duplexes. The result is that the Cu(II) line up in the center of the double helix (Fig. 10). In addition, the presence of Cu^{2+} will reduce the electrostatic repulsion significantly. For example, the thermal stability of the DNA-Cu conjugate is enhanced as a result. Shionoya and coworkers reported that a sequential sequestering of the Cu ions by the hydroxypyridone bases. The metal ions are lined up in the center of the right-handed helix with a Cu–Cu distance of about 3.7 ± 0.1 Å, which was estimated from molecular modeling. EPR studies unequivocally demonstrate that the Cu ions are electronically coupled. The close proximity within the confines of the double helix allows for a parallel alignment of the unpaired spins (Cu^{2+} is d^9) to give ferromagnetic behavior of the metallated DNA helix. This result is significant in that it suggests that the Cu-metallated DNA may act as an effective molecular wire, allowing signal transmission between either ends of the double helix. However, at the present time, no data are available that unequivocally demonstrate this to be the case.

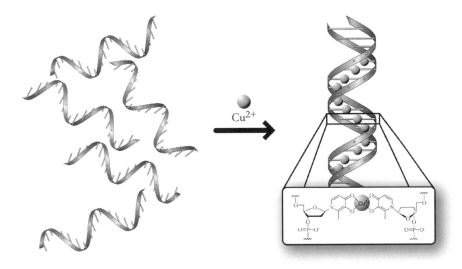

Figure 10 Cu^{2+} incorporation into hydroxypyridone nucleotide-labeled DNA forms Cu complexes in which the Cu occupies the center position in the DNA duplex (redrawn from Ref. 32).

B. M-DNA

A novel method for improving the conductivity of DNA involves close association or even incorporation of divalent metal ions such as Zn^{2+}, Ni^{2+}, and Co^{2+} into the helix at pH>8.5 to form metallated DNA (M-DNA).[33] Other divalent metals such as Mg^{2+} and Ca^{2+} do not form M-DNA. In photophysical studies using ds-DNA where one strand is labeled with a fluorescein fluorophore and the complementary strand carries a rhodamine quencher, the fluorescein fluorescence is quenched only under conditions where M-DNA forms. M-DNA is an efficient conductor of electrons over distances as long as 500 basepairs and possibly as long as several micrometers. It was proposed that in M-DNA the Zn^{2+} cations replace the imino protons of every thymine and guanine in addition to saturating the backbone phosphates as do other cations such as Mg^{2+} (Scheme 6).[33b] Importantly, M-DNA can be converted back to ordinary ds-DNA by the addition of ethylenediaminetetraacetic acid (EDTA), which will sequester the metal ions.

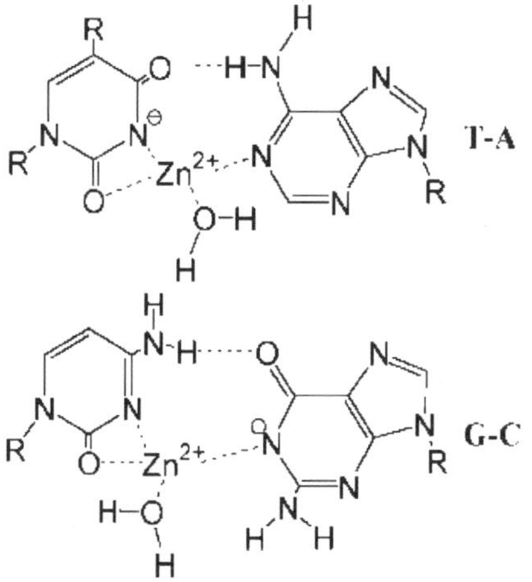

Scheme 6 Possible sites of incorporation of divalent metal ions, such as Zn^{2+}, into basepairs.

A schematic view of the metal incorporation into the DNA is given in Scheme 7.

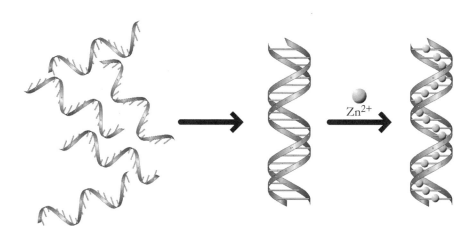

Scheme 7 Conversion of ds-DNA (B-DNA) to M-DNA in the presence of Zn^{2+} at pH >8.5.

Kraatz, Lee and coworkers demonstrated that M-DNA also forms in the confines of ds-DNA bound to a gold surface.[34]

Utilizing the 5'-disulfide-modified DNA construct HO-$(CH_2)_6$SS-$(CH_2)_6$5'-GTCACGATGGCCCAGTAGTT-3' and its complementary strand 5'-AAC-TACTGGGCCATCGTGAC-3', a ds-DNA monolayer was formed as shown in Scheme 8. In the presence of Zn^{2+} at pH 8.6, the surface slowly converts to an M-DNA surface. Figure 11 shows a time study on the effects of the addition of Zn^{2+} and Mg^{2+}

on the CV of ds-DNA/hydroxyalkyl monolayer in the presence of $[Fe(CN)_6]^{3-/4-}$ as the solution electrophore. In this format, electron transfer has to proceed from the hexacyanoferrate(II) to the electrode surface via the DNA film. Before the addition of any metal ions, the ds-DNA monolayer blocks the ET process effectively, indicating only poor ET through the DNA film. However, after the addition of Zn^{2+} to the buffer solution, the signal intensity of the CV of the $[Fe(CN)_6]^{3-/4-}$ electrophore in solution steadily increases up to a limit and the peak separation decreases. Importantly, for Zn^{2+}, the peak current increases significantly and then reaches a maximum after about 2 h, and then remains at this maximum, corresponding to about 80% of the signal intensity of bare gold. In contrast, for Mg^{2+}, the increase over the same period is comparatively small. Thus, the electrochemical properties of the ds-DNA/hydroxyalkyl monolayer in the presence of Mg^{2+} are not affected.

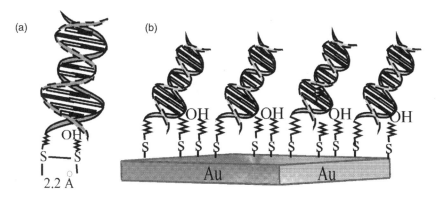

Scheme 8 Schematic representation of the DNA construct before and after attachement onto the gold surface. Note the formation of a mixed monolayer consisting of 6-hydroxyhexanolthiolate and DNA-hexanolthiolate. This will allow some spacing between the DNA molecules on the surface.

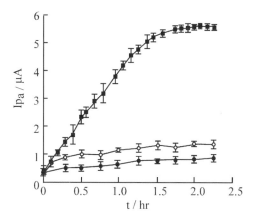

Figure 11 Anodic peak current as function of incubation time for ds-DNA films on gold electrodes in (●) 2.5 mM $[Fe(CN)_6]^{3-/4-}$ in Tris-ClO$_4$ buffer solution (20 mM, pH 8.6) with the addition of (■) 0.3 mM Zn(ClO$_4$)$_2$; (○) 0.3 mM MgCl$_2$. The error bars represent the standard deviation on three independent determinations.

The time-dependent increase of the electrical response probably represents the dynamics of incorporation of Zn^{2+} into the ds-DNA monolayer. To a first approximation, the M-DNA formation on the surface follows a pseudo-first-order kinetics with a pseudo-first-order rate constant of $k=10$ min^{-1}.

As mentioned above, the addition of the chelator EDTA to a buffer solution with an M-DNA electrode will complex to the Zn^{2+} causing the conversion of M-DNA back into native ds-DNA, concomitant with a loss of the particular electronic properties of M-DNA. Figure 12 shows the CV curve of M-DNA in the absence and presence of EDTA. On addition of EDTA to the buffer containing the M-DNA monolayer, the signal intensity is significantly decreased and the peak separation ΔE_p increases, indicating a loss of signal transduction from the redox probe through the monolayer after removal of the Zn^{2+} ions by EDTA. The resulting CV curve exhibits the same features as that of a native ds-DNA film. Thus, we conclude that, as in solution, M-DNA can be converted back to native ds-DNA on the gold surface.

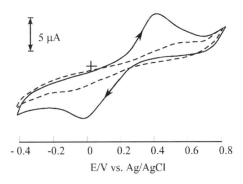

Figure 12 Cyclic voltammograms of M-DNA-modified gold electrodes (——) before and after treatment with 1 mM EDTA (-----) (2.5 mM $[Fe(CN)_6]^{3-/4-}$ in 20 mM Tris-ClO$_4$ buffer (pH 8.6) at a sweep rate of 100 mV/s).

In addition, the heterogeneous ET kinetics for bare gold and the M-DNA modified electrode were obtained by comparing the experimental with a simulated CV curve. Figure 13 shows the CV curves for bare Au electrode (Fig. 13a) and the M-DNA SAM-modified Au electrode (Fig. 13b), together with the simulated CV plot.

Under the experimental conditions, the rate constant k_{ET} for the bare gold electrode was $9.1\times10^{-4}\pm0.2\times10^{-4}$ cm/s and $1.2\times10^{-4}\pm0.2\times10^{-4}$ cm/s for the M-DNA/hydroxylalkyl-modified Au electrode. Interestingly, for ds-DNA/hydroxylalkyl-modified Au electrode, k_{ET} was too small to be measured. This indicates that k_{ET} for native ds-DNA must be orders of magnitude smaller! These results were confirmed by chronocoulometric measurements on the two types of DNA films.

AC impedance spectroscopy (ACIS) is extremely useful to evaluate the electronic properties of monolayers. Using an equivalence circuit, the impedance of the system can be expressed in terms of simple circuit elements, such as capacitors, C, and resistors, R. The impedance spectra of ds-DNA and M-DNA together with the equivalent circuit was investigated.[35] ACIS measurements were performed in the presence of 4 mM $[Fe(CN)_6]^{3-/4-}$ (1–1) mixture, as the solution-based redox probe. Figure 14 shows a Nyquist plot for a 20-mer ds-DNA on gold. The diameter of the semicircle is a measure of the charge transfer resistance, R_{ct}. The data were fit to a modified Randles circuit having an additional resistor in parallel, R_x, which gives an excellent fit to the experimental data in both the low- and high-frequency zones. Addition of Zn^{2+} to the 20-mer ds-DNA modified gold electrode at pH 8.7 results in the formation of M-DNA as described above.[34]

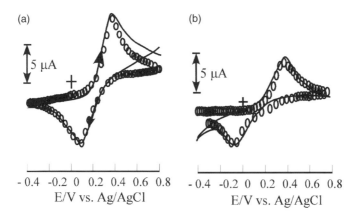

(a)   (b)

5 µA   5 µA

- 0.4 -0.2 0 0.2 0.4 0.6 0.8
E/V vs. Ag/AgCl

- 0.4 -0.2 0 0.2 0.4 0.6 0.8
E/V vs. Ag/AgCl

Figure 13 Cyclic voltammograms(s) of 2.5 mM $[Fe(CN)_6]^{3-/4-}$ in 20 mM Tris-ClO$_4$ (pH 8.6) at (a) a bare gold electrode and (b) a M-DNA-modified gold electrode. Simulated data are indicated as (O) (scan rate = 100 mV/s. Digital simulations were made using the k_{ET}, $E°$: (a) 9.1×10^{-4}, 0.21 (vs. Ag/AgCl), and $\alpha = 0.5$; (b) 1.23×10^{-4} cm/s, 0.19 V (vs. Ag/AgCl), and $\alpha = 0.4$. The diffusion coefficients of D_{ox} and D_{red} used for (a) were 8×10^{-6} cm^2/s for the $[Fe(CN)_6]^{3-/4-}$ couple.

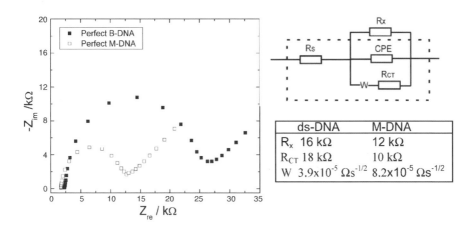

	ds-DNA	M-DNA
R_x	16 kΩ	12 kΩ
R_{CT}	18 kΩ	10 kΩ
W	3.9×10^{-5} Ωs$^{-1/2}$	8.2×10^{-5} Ωs$^{-1/2}$

Figure 14 Nyquist plots of a 20-mer of ds-DNA (■) and M-DNA (□). The spectra were fit to the equivalent circuit shown on the top left; individual elements of the equivalent circuit are summarized below: charge transfer resistance R_{CT}, Warburg impedance W, and interfacial resistance R_x.

The impedance spectrum undergoes a significant change with a reduction in Z_{im} and Z_{re} at both high and low frequencies. The data were fit with the identical equivalent circuit. On M-DNA formation, R_{CT} and R_x decrease significantly. In the presence of other metals, such as Mg^{2+} or in a region of the pH where M-DNA does not form (below pH 8.5), no changes in the impedance were observed, indicating that the changes are metal- and pH-specific. Importantly, the effects scale with the length of the DNA. There are two distinct trends. First, R_x and R_{CT} increase with increasing length for both ds- and M-DNA. The distance dependence is shallow, pointing to a hopping mechanism for the ET process, as was originally proposed by Giese.[3a] Second, for any length of duplex R_x and R_{CT} for M-DNA is less than the corresponding value for ds-DNA. The Warburg impedance W, which represents mass transfer to the electrode, is more variable but in all cases is higher for the M-DNA duplexes, most likely due to the presence of additional metal ions in solution. As expected, the double-layer capacitance C decreases with increasing length of the duplex. These results indicate that M-DNA is a better conductor than native ds-DNA since both R_{CT} and R_x are smaller for M-DNA. These results compare well with Shionoya's results[32] of Cu-DNA. In both cases, the tight interaction of the metal with the DNA double helix is expected to enhance the electronic communication significantly.

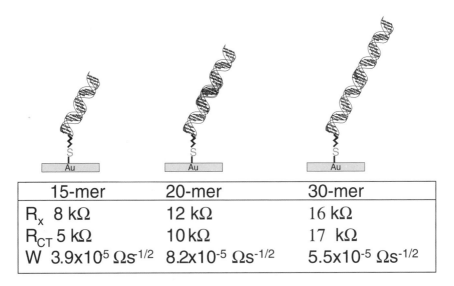

15-mer	20-mer	30-mer
R_x 8 kΩ	12 kΩ	16 kΩ
R_{CT} 5 kΩ	10 kΩ	17 kΩ
W 3.9x10^5 Ωs$^{1/2}$	8.2x10^{-5} Ωs$^{-1/2}$	5.5x10^{-5} Ωs$^{-1/2}$

Figure 15 Length dependence of the resistance R_{CT} and R_X.

Thus the question arises as to whether it is possible to detect a break in the M-DNA, as may be caused by the presence of a basepair mismatch. Early results on the electrochemical impedance behavior of mismatched DNA films suggest that mismatched DNA films behave differently from fully matched DNA films.

VI. SUMMARY

The self-assembly properties of DNA have been exploited to generate 2D and 3D structures on surfaces and in solution. DNA now can be modified by metal centers at individual bases or at either termini. The ability to label DNA site specifically with metal centers allows influence of the electronic properties of the assembly. Potential applications of these modified DNA constructs in the design of nanoelectronic or bioelectronic circuitry may be within reach. Especially the most recent developments in the area of direct basepair metallation, such as Cu-DNA and M-DNA, allow superb control over the electronics of the DNA construct and reduce the synthetic efforts.

As shown in this review, a large driving force for research in the area of metal-DNA conjugates stems from sensor applications, for example, in the detection of genetic defects, genomic fingerprinting for identification purposes, or applications in personalized medicine. Electrochemical detection methods offer a superior sensitivity to currently available optical gene chip technology. Gene chip technology has to rely on attaching a fluorescence tag to one of the DNA strands after PCR amplification. More recent advances in the area of electrochemical DNA detection described in this review should allow the development of biosensors that do not require labeling of the target DNA.

One of the holy grails of E-biosensors is the detection of single nucleotide mismatches time- and cost-effectively without the need for a lengthy PCR amplification, or in heterozygote mixtures or under nonideal hybridization conditions. These goals are yet unrealized. Especially difficult will be the detection of DNA under real-life conditions, such as high salt concentrations or in the presence of large quantities of impurities. However, the possibility of single molecule detection using ultra fast electrochemical techniques on ultramicroelectrodes may offer solutions to this problem.

VII. ACKNOWLEDGMENTS

This work was supported by the National Science and Engineering Research Council. HBK is Canada Research Chair in Biomaterials.

VIII. REFERENCES

1. (a) R. Elghanian, J. J. Storhoff, R. C. Mucic, R. L. Letsinger, C. A. Mirkin, *Science* **277,** 1078 (1997); (b) B. H. Bobinson, N. C. Seeman, *Protein Eng.* **1,** 295 (1987); (c) J. J. Gooding, D. B. Hibbert, W. Yang, *Sensors* **1,** 75 (2001); (d) C. M. Niemeyer, *Appl. Phys.* **68,** 119 (1999).

2. (a) M. Bixon, J. Jortner, *J. Am. Chem. Soc.* **123,** 12556 (2001); (b) M. Bixon, B. Giese, S. Wessely, T. Langenbacher, M. E. Michel-Beyerle, J. Jortner, *Proc. Natl. Acad. Sci. USA.* **96,** 11713 (1999); (c) P. T. Henderson, D. Jones, G. Hampikian, Y. Z. Kan, G. B. Schuster, *Proc. Natl. Acad. Sci. USA.* **96,** 8353 (1999); (d) R. N. Barnett, C. L. Cleveland, A. Joy, U. Landman, G. B. Schuster, *Science* **294,** 567 (2001).

3. (a) E. M. Boon, J. K. Barton, *Curr. Opin. Struct. Biol.* **12,** 320 (2002); (b) B. Giese, *Curr. Opin. Chem. Biol.* **6,** 612 (2002). (c) G. B. Schuster, *Acc. Chem. Res.* **33,** 253 (2000); (d) S. Delaney, J. K. Barton, *J. Org. Chem.* **68,** 6475 (2003).

4. (a) E. K. Wilson *C&EN* **24**, 33 (1997); b) C. J. Murphy, M. R. Arkin, Y. Jenkins, N. D. Ghatlia, S. H. Bossmann, N. J. Turro, J. K. Barton, *Science* **262**, 1025 (1993).

5. (a) A. Anne, B. Blanc, J. Moiroux, *Bioconj. Chem.* **12**, 396 (2001); (b) R. C. Mucic, M. K. Herrlein, C. A. Mirkin, R. L. Letsinger, *Chem. Commun.* 555 (1999); (c) C. J. Yu, H. Wang, Y. J. Wan, H. Yowanto, J. C. Kim, L. H. Donilon, C. L. Tao, M. Strong, Y. C. Chong, *J. Org. Chem.* **66**, 2937 (2001); (d) M. T. Tierney, M. W. Grinstaff, *J. Org. Chem.* **65**, 5355 (2000); (e) K. Yamana, S. Kumamoto, T. Hasegawa, H. Nakano, Y. Sugie, *Chem. Lett.* 506 (2002).

6. H.-B. Kraatz, J. Lusztyk, G. D. Enright, *Inorg. Chem.* **36**, 2400 (1997).

7. A. R. Pike, L. C. Ryder, B. R. Horrocks, W. Clegg, M. R. J. Elsegood, B. A. Connolly, A. Houlton, *Chem. Eur. J.* **8**, 2891 (2002).

8. T. Ihara, M. Nakayama, M. Murata, K. Nakano, M. Maeda, *Chem. Commun.* 1609 (1997).

9. M. Nakayama, T. Ihara, K. Nakano, M. Maeda, *Tantala* **56**, 857 (2002).

10. F. Patoslky, Y. Weizmann, I. Willner, *J. Am. Chem. Soc.* **124**, 770 (2002).

11. A. E. Beilstein, M. W. Grinstaff, *Chem. Commun.* 509 (2000).

12. W. A. Wlassoff, G. C. King, *Nucl. Acid Res.* **30**, e58 (2002).

13. (a) C. Fan, K. W. Plaxco, A. J. Heeger, *Proc. Natl. Acad. Sci. USA* **100**, 9134 (2003); (b) X. Hu, Ph.D. thesis, Duke Univ., 2001.

14. R. M. Umek, S. W. Lin, J. Vielmetter, R. H. Terbrueggen, B. Irvine, C. J. Yu, J. F. Kayyem, H. Yowanto, G. F. Blackburn, D. H. Farkas, Y.-P. Chen, *J. Mol. Diagnostics* **3**, 74 (2001).

15. M. A. O'Neill, J. K. Barton, *Proc. Natl. Acad. Sci. USA* **99**, 16543 (2002).

16. Y.-T. Long, C.-Z. Li, T. C. Sutherland, M. Chahma, J. S. Lee, H.-B. Kraatz, *J. Am. Chem. Soc.* **125**, 8724 (2003).

17. (a) P. E. Laibinis, G. M. Whitesides, *J. Am. Chem. Soc.* **114**, 1990 (1992); (b) R. G. Nuzzo, L. H. Dubois, D. L. Allara, *J. Am. Chem. Soc.* **112**, 558 (1990); (c) C. D. Bain , E. B. Troughton, Y. -T. Tao, J. Evall, G. M. Whitesides, R. G. Nuzzo, *J. Am. Chem. Soc.* **111**, 321 (1989).

18. (a) P. E. Laibinis, C. D. Bain, G. M. Whitesides, *J. Phys. Chem.* **95**, 7017 (1991); (b) P. E. Laibinis, G. M. Whitesides, D. L. Allara, Y.-T. Tao, A. N. Parikh, R. G. Nuzzo, *J. Am. Chem. Soc.* **113**, 7152 (1991).

19. A. J. Bard, L. R. Faulkner, *Electrochemical Method: Fundamentals and Applications*; 2nd ed., Wiley, New York, 2001.

20 S. A. Brazill, P. H. Kim, W. G. Kuhr, *Anal. Chem.* **73**, 4882 (2001).

21. C. Xu, P. He, Y. Fang, *Anal. Chim. Acta* **411**, 31 (2000).

22 A. Hess, N. Metzler-Nolte, *Chem. Commun.* 885 (1999).

23. K. Plumb, H.-B. Kraatz, *Bioconj. Chem.* **14**, 601 (2003).

24. T. J. Meade, J. F. Kayyem, *Angew. Chem. Int. Ed.* **34**, 352 (1995).

25. J. J. Rack, E. S. Krider, T. J. Meade, *J. Am. Chem. Soc.* **122**, 6287 (2000).

26. N. L. Frank, T. J. Meade, *Inorg. Chem.* **42**, 1039 (2003).

27. S. I. Khan, A. E. Beilstein, M. Sykora, G.D. Smith, X. Hu, M. W. Grinstaff, *Inorg. Chem.* **38**, 3922 (1999).

28. D. J. Hurley, Y. Tor, *J. Am. Chem. Soc.* **124**, 3749 (2002).

29. R. E. Holmlin, P. J. Dandliker, J. K. Barton, *Bioconj. Chem.* **10**, 1122 (1999).

30. (a) M. R. Arkin, E. D. A. Stemp, R. E. Holmlin, J. K. Barton, A. Hormann, E. J. C. Olsen, *Science* **273**, 475 (1996); (b) M. R. Arkin, E. D. A. Stemp, C. Turro, N. J. Turro, J. K. Barton, *J. Am. Chem. Soc.* **118**, 2267 (1996).

31. K. Kowalski, J. Zakrzewski, *J. Organomet. Chem.* **2003**, *668*, 91.

32. K. Tanaka, A. Tengeiji, T. Kato, N. Toyama, M. Shionoya, *Science* **299**, 1212 (2003).

33. (a) P. Aich, S. L. Labiuk, L. W. Tari, L. J. T. Delbaere, W. J. Roesler, K. J. Falk, R. P. Steer, J. S. Lee, *J. Mol. Biol.* **294**, 477 (1999); (b) J. S. Lee, L. J. P. Latimer, R. S. Reid, *Biochem. Cell Biol.* 162 (1993).

34. C.-Z. Li, Y.-T. Long, H.-B. Kraatz, J. S. Lee, *J. Phys. Chem. B* **107**, 2291 (2003).

35. Y.-T. Long, C.-Z. Li, H.-B. Kraatz, J. S. Lee, *Biophys. J.* **84**, 3218 (2003).

36. Y.-T. Long, C.-Z. Li, T. C. Sutherland, H.-B. Kraatz, J. S. Lee, unpublished results.

CHAPTER 3

Artificial DNA through Metal-Mediated Base Pairing: Structural Control and Discrete Metal Assembly

Mitsuhiko Shionoya

Department of Chemistry, Graduate School of Science,
The University of Tokyo, Tokyo, Japan

CONTENTS

I. INTRODUCTION   46

II. ALTERNATIVE HYDROGEN-BONDING SCHEMES FOR DNA
BASE PAIRING   46

III. NON-HYDROGEN-BONDING BASE PAIRS IN DNA   48

IV. METAL-MEDIATED BASE PAIRING IN DNA   49
    A. Basic Concept   49
    B. Artificial Nucleosides Designed for Metal-Mediated Base Pairs   49
    C. Incorporation of a Metallo-Base Pair in DNA and Its Effect on Thermal
       Stability   50
    D. Discrete Self-Assembled Metal Arrays in DNA   52

V. FUTURE PROSPECTS FOR ARTIFICIAL METALLO-DNA   54

Macromolecules Containing Metal and Metal-Like Elements,
Volume 3: Biomedical Applications, edited by Alaa S. Abd-El-Aziz,
Charles E. Carraher Jr., Charles U. Pittman Jr., John E. Sheats, and Martel Zeldin
ISBN: 0-471-66737-4   Copyright © 2004 John Wiley & Sons, Inc.

VI. SUMMARY   54

VII. REFERENCES   55

I. INTRODUCTION

Research on biologically inspired molecular architecture is often motivated by the belief that "bottom–up" approaches to renew fundamental building blocks that have been provided by Nature can lead to a wide range of possible structures and functions of the final architectures. Self-assembly protocols and template-directed procedures efficiently used in the biological systems have been conceptually introduced into nonbiological approaches to self-assembled, nanosized molecules or materials. However, the biologically related aspects of molecular architecture are, although promising, not as well advanced as the nonbiological ones. Although biopolymers contain only a limited number of building blocks such as nucleotides and amino acids, more recent developments in chemical synthesis and biotechnology allow one to replace the building blocks by chemically reconstructed ones and arrange them one after another with a desired length and sequence. In addition, it has been generally accepted that the incorporation of metal complexes as alternative components into biomolecular scaffolds is a key design in the structural control and functionalization of biopolymers.

DNA has a structural basis to arrange functionalized building blocks into pre-designed geometries. In the double helix formed from two complementary oligonucleotide strands, hydrogen-bonded DNA base pairs, which are attached nearly perpendicular to the phosphate backbone, are arranged into direct-stacked contact (Fig. 1). Therefore, among a variety of approaches to DNA-based supramolecular architectures, the strategy of replacing natural DNA base pairs by alternative ones possessing a distinctive shape, size, and function[1] is expected to provide a general method of controllable molecular arrangement within the DNA. Such extra base pairs would not only expand the genetic alphabet but also allow the replication of DNA containing unique functional groups. Moreover, DNA that is completely built out of artificial base pairs could lead to novel oligomers or polymers having interesting chemical and physical properties.

This review covers recent advances in the replacement of natural DNA bases by artificial ones directed toward DNA nanotechnology as well as gene control.

II. ALTERNATIVE HYDROGEN-BONDING SCHEMES FOR DNA BASE PAIRING

Watson–Crick hydrogen bonding in natural base pairs is essential for most of the DNA functions. Initial effort was made largely to expand the genetic alphabet using altered hydrogen bonding. Benner and coworkers pioneered a way to enzymatic

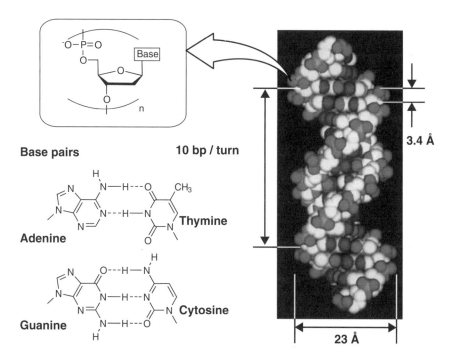

Base pairs   **10 bp / turn**

3.4 Å

23 Å

Figure 1 A-T and G-C Watson–Crick base pairs found in natural DNA.

incorporation of new hydrogen-bonded base pairs into DNA and RNA to extend the genetic alphabet.[2a] They reported a series of nucleobase analogs whose hydrogen-bonded patterns differ from those in the adenine–thymine (A-T) and guanine–cytosine (G-C) base pairs of natural DNA. Considering hydrogen-bonded patterns capable of forming Watson–Crick base pairs with three hydrogen bonds, there are at least six mutually exclusive hydrogen-bonding schemes. Interestingly, even these subtle changes in their hydrogen-bonding patterns have a great influence on thermodynamic[2b] and biochemical properties.[2c–e] For example, isoguanine (**iso-G**) and isocytosine (**iso-C**) can form a Watson-Crick basepair with a standard geometry (Fig. 2), in which the hydrogen-bonding pattern is different from those found in the natural base pairs A-T and G-C. Interestingly, several polymerases were found to incorporate

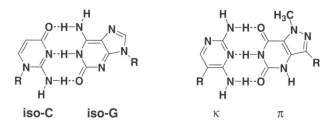

iso-C   iso-G   κ   π

Figure 2 Examples of base pairs with alternative hydrogen-bonding schemes.

iso-G opposite **iso-C** in a template.[2d] Furthermore, an additional set of codon–anti-codon [((**iso-C**)AG)-(CU(**iso-G**))] was created by the (**iso-G**)–(**iso-C**) base pair. This codon-anticodon set was successfully used to incorporate an unnatural amino acid, iodotyrosine, into a polypeptide prepared by a ribosome through translation of a messenger RNA containing the additional codon.[2c]

III. NON-HYDROGEN-BONDING BASE PAIRS IN DNA

Watson–Crick base pairing in complementary oligonucleotide strands keeps two rules of complementarity in both size and hydrogen-bonding patterns. Hydrophobicity and planarity in the bases also appear to be important for the stability of the double-helical structure. Designing new base pairs that vary in shape, size, and functionality has been useful in understanding what is essential in the natural base pairing.

Along this line, Kool and coworkers have developed a series of shape mimics of natural bases lacking hydrogen bonding functionality, using the principle that two bases must be complementary in shape rather than in hydrogen bonding.[3] A set of non-hydrogen-bonding base mimics for thymine and adenine is shown in Figure 3. The design criteria are to replace oxygen with fluorine and nitrogen with carbon and to keep aromaticity intact. Because of the lack of polar functional groups, the final structures are nonpolar and hydrophobic. For example, the nucleoside bearing a difluorotoluene nucleobase (**F**) is an excellent mimic of thymidine in the crystalline state and in solution.[3d] Even when **F** is paired with adenine within DNAs, it forms a structure very similar to that of a T-A base pair.[3e] Moreover, a benzimidazole isostere (**Z**) of adenine base acts as an outperforming mimic in DNAs.[3b,f] Despite the lack of hydrogen-bonding functionality, some shape mimics of DNA bases have been found to retain normal DNA functions. For example, **F** can effectively substitute for thymine as the incoming substrate in the triphosphate form[4] as well as in the template strand[5a] in polymerase-related enzymatic reactions. Especially in the latter case when **F** was in a template strand of DNA, despite the fact that **F** pairs equally well with all four bases, the Klenow fragment of *E. coli* DNA Pol I could efficiently and selectively incorporate adenine opposite **F**. These results overall suggest that enzymatic replication of base pairs does not need Watson–Crick hydrogen bonds as long

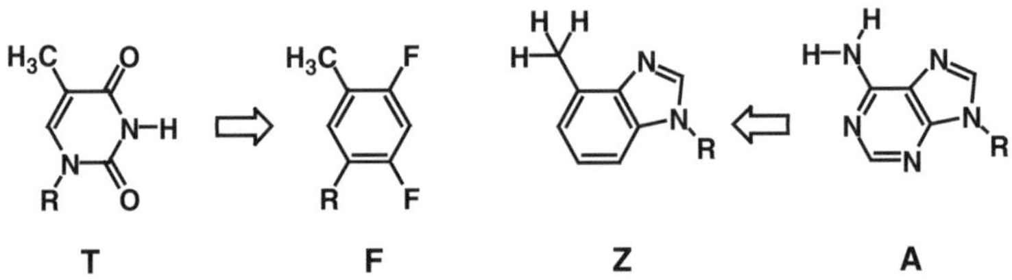

Figure 3 Examples of nonpolar isosteres for T-A base pair, **F-Z**.

as the components stack strongly. They also suggested that shape recognition is important in the base pairing without hydrogen bonding.[5b] Others have also reported hydrophobic unnatural base pairs as attractive candidates for expansion of the genetic alphabets.[6]

IV. METAL-MEDIATED BASE PAIRING IN DNA

A. Basic Concept

When natural DNA bases are replaced by unnatural ones that have the ability to bind metal ions, metal-mediated base pairs could possibly be incorporated into DNAs at predesigned positions or aligned along the helix axis. Metal coordinative bond energy (10–30 kcal/mol) is intermediate between covalent and noncovalent ones. Accordingly, one metal–ligand bond should compensate for two or three hydrogen bonds as found in the natural base pairs. Metal ions incorporated in this way could, for instance, possibly (1) regulate thermal stability of high-order structures of DNA (duplex, triplex, etc.), (2) allow one-dimensional metal arrays (or strings, if in direct-stacked contact) along the DNA helix axis with interesting chemical and physical properties, (3) generate metal-dependent functions such as electro- or photochemical catalysts, (4) assemble DNA duplexes at the junctions to form DNA two- or three-dimensional networks, or (5) label DNA with metal ions at desired positions. Some of these possibilities have been supported by the literature.

B. Artificial Nucleosides Designed for Metal-Mediated Base Pairs

In 1999, we described the first artificial metal ligand-type nucleoside having a phenylenediamine base for Pd^{2+}-mediated base pairing[7] as an approach to the replacement of natural base pairs by alternative ones. Since then, a few β-*N*- and β-*C*-nucleosides for metal-mediated base pairing were synthesized by other groups. Examples of artificial nucleosides having mono- to tridentate ligands for metals are shown in Figure 4. Each nucleobase has one to three donor atoms at proper positions so that the ligand moiety can form a 2:1, 3:1, or 4:1 complex with a transition metal ion in a linear, trigonal–planar, square–planar, tetrahedral, or octahedral manner. Among these coordination geometries, a square–planar, linear, or trigonal-planar metal complex can replace a flat, hydrogen-bonded natural base pair. So far, we have reported, in addition to the abovementioned Pd^{2+}-mediated base pairing, B^{3+}-induced base paring with catechol,[8] Pd^{2+}-mediated base pairing with 2-aminophenol,[9] Ag^+-assisted base pairing with pyridine,[10] and Cu^{2+}-mediated base pairing with hydroxypyridone[11] as alternative base pairing modes (Fig. 5). Other groups have also reported metal-mediated association of ligand nucleobase mimics such as a Cu^{2+}-mediated [1+3]-type base pair between pyridine and pyridine-2,6-dicarboxylate,[12a,b]

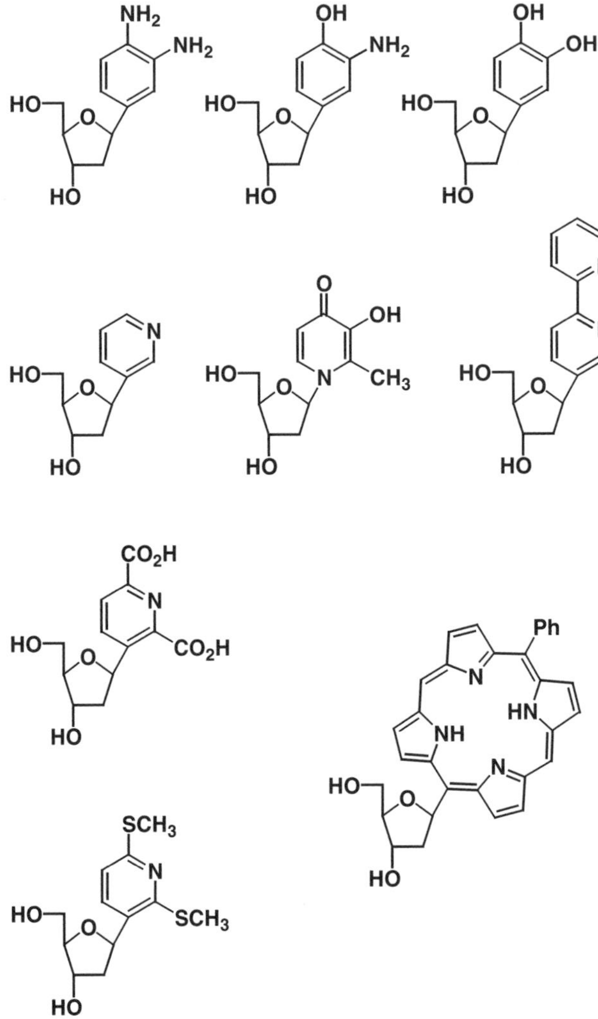

Figure 4 Examples of artificial nucleosides in which DNA bases are replaced by ligands for metal ions.

a Ag$^+$-mediated one with 2,6-bis(methylthiomethyl)pyridine bases,[12c] and "ligando-sides" using bipyridine nucleobases.[13]

C. Incorporation of a Metallo-Base Pair in DNA and Its Effect on Thermal Stability

Incorporation of artificial nucleobases into oligodeoxynucleotides using phos-phoramidite derivatives of the nucleosides can be carried out with standard proto-cols using an automated DNA synthesizer. The first example of metal-mediated

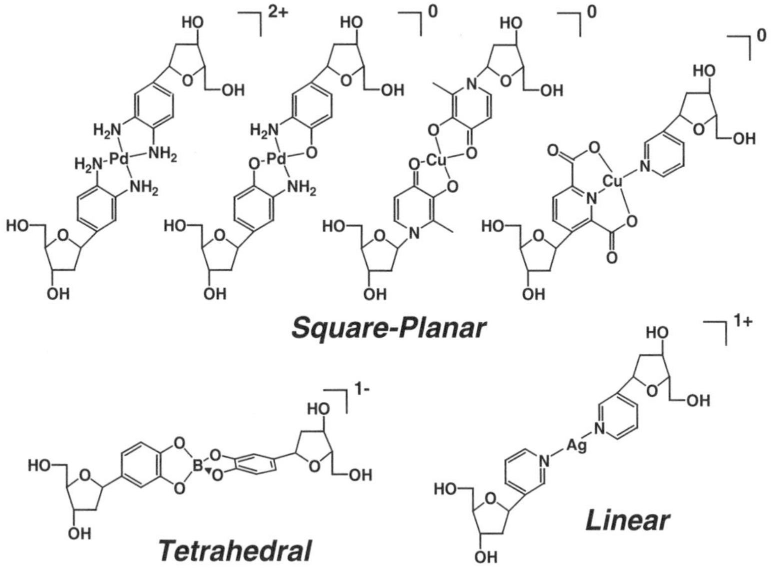

Figure 5 Examples of metal-mediated base pairs through metal-coordinating nucleoside analogs.

base pairing in oligonucleotides was described by Schultz and coworkers[12a] using a base pair in the middle of a 15-nucleotide duplex with a pyridine-2,6-dicarboxylate nucleobase as a planar tridentate ligand, and a pyridine nucleobase as the complementary single donor ligand (Fig. 5). The duplex is significantly stabilized by Cu^{2+} ions due to the formation of a neutral Cu^{2+} complex with the paired ligand bases inside the DNA, and the stability is comparable to that of a duplex containing an A-T base pair instead of the Cu^{2+}-mediated one.

We have independently established a Ag^+-mediated base pair in a double-stranded DNA by introducing a monodentate pyridine nucleobase in the middle of each strand.[10] For example, a Ag^+ ion incorporated into a DNA duplex, $d(5'\text{-}T_{10}PT_{10}\text{-}3')\cdot d(3'\text{-}A_{10}PA_{10}\text{-}5')$, containing a pyridine nucleobase (**P**) in the middle of the sequence, increases the thermal stability of the duplex, due to the positively charged **P**-Ag^+-**P** base pairing. In contrast, this Ag^+-dependent thermal stabilization of duplex is only slight in a reference DNA duplex, $d(5'\text{-}T_{21}\text{-}3')\cdot d(3'\text{-}A_{21}\text{-}5')$. It is to be noted that the **P**-Ag^+-**P** base pair can be formed inside the DNA even at the concentrations of micromolar order in spite of the relatively weak binding constant of pyridine with Ag^+ in aqueous media. The complex formation of Ag^+ and pyridine in the artificial DNA should be reinforced cooperatively by surrounding hydrogen-bonded and stacked base pairs in the hydrophobic environment within the duplex. Moreover, this Ag^+-mediated base pairing is specific because the addition of other transition metal ions such as Cu^{2+}, Ni^{2+}, Pd^{2+}, and Hg^{2+} shows almost no significant effects. Such thermal stabilization is also observed with a triplex, $d(5'\text{-}T_{10}PT_{10}\text{-}3')\cdot d(3'\text{-}A_{10}PA_{10}\text{-}5')\cdot d(5'\text{-}T_{10}PT_{10}\text{-}3')$.[10] This effect is believed to be due to the

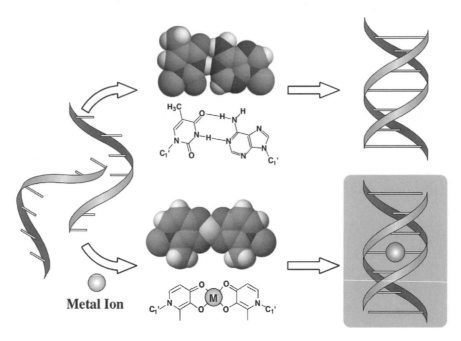

Figure 6 A schematic representation for single-site incorporation of a metal-mediated base pair.

formation of a Ag^+-mediated base triplet in which the nitrogen donors of the three pyridine nucleobases coordinate to the Ag^+ center.

In this regard, a hydroxypyridone-bearing nucleoside forms a neutral Cu^{2+}-mediated base pair of hydroxypyridone nucleobases ($\mathbf{H}$-Cu^{2+}-$\mathbf{H}$) and can be incorporated into a 15-nucleotide DNA duplex, $d(5'$-CACATTA**H**TGTTGTA-$3')$ $\cdot d(3'$-GTGTAAT**H**ACAACAT-$5')$.[11] In the presence of equimolar Cu^{2+} ions, an $\mathbf{H}$-Cu^{2+}-$\mathbf{H}$ base pair is quantitatively formed within the DNA (see Fig. 6) and the artificial duplex showed a higher thermal stability compared with a natural oligoduplex, $d(5'$-CACATTA**A**TGTTGTA-$3')\cdot d(3'$-GTGTAA**T**TACAACAT-$5')$, in which the $\mathbf{H}$-$\mathbf{H}$ base pair is replaced by an $\mathbf{A}$-$\mathbf{T}$ base pair. In addition, EPR and CD spectra of the metallo-DNA suggest that the radical site of a Cu^{2+} center is formed within the right-handed double-strand structure of the oligonucleotide. This strategy was further developed for controlled and periodic spacing of neutral metallo-base pairs along the helix axis of DNA.

D. Discrete Self-Assembled Metal Arrays in DNA

To control metal arrays in a discrete and predictable manner, one definitely needs appropriate ligands that have a varied number of coordination sites. From this perspective, DNA is a promising molecule that could act as a multidentate ligand for one-dimensional metal arrays when the nucleobases are replaced by ligand-like bases. Thus, incorporation and the subsequent arrangement of metallo-base pairs into direct-stacked contact within DNA could lead to "metallo-DNA" in which metal ions are lined up along the helix axis in a controlled and stepwise manner.

The synthesis of a series of artificial oligonucleotides, $d(5'\text{-}GH_nC\text{-}3')$ ($n = 1$–5), was reported using hydroxypyridone nucleobases (**H**) as flat bidentate ligands.[14] Right-handed double helices of the oligonucleotides, $n\mathrm{Cu}^{2+}{\cdot}d(5'\text{-}GH_nC\text{-}3')_2$ ($n = 1$–5), are quantitatively formed through Cu^{2+}-mediated alternative base pairing (**H-Cu^{2+}-H**) (Fig. 7). In these metallo-DNA, the Cu^{2+} ions incorporated into each complex are aligned along the helix axes inside the duplexes with a Cu^{2+}–Cu^{2+} distance of 3.7 ± 0.1 Å. The Cu^{2+} ions are coupled with one another through unpaired d electrons to form magnetic chains. The electron spins on adjacent Cu^{2+} centers are aligned parallel and

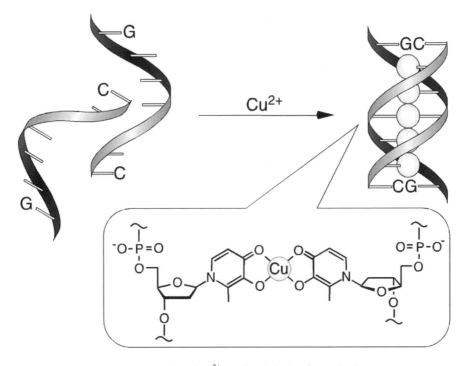

Figure 7 Schematic representation of Cu^{2+}-mediated duplex formation between two artificial DNA strands in which hydroxypyridone nucleobases replace natural base pairs.

couple in a ferromagnetic fashion with accumulating Cu^{2+} ions attaining the highest spin state, as expected from a lineup of Cu^{2+} ions. An outline of a proposed right-handed and double-stranded structure is drawn with Cu^{2+} array inside the DNA (Fig. 8).

Such strategy could be developed for self-assembled metal arrays in a variety of DNA structures such as multistranded, hairpin, junction, or cyclic structures. The next approach should be to increase the number of metal ions or to create heterometal arrays possibly leading to metal–metal communication triggered by stimuli from outside.

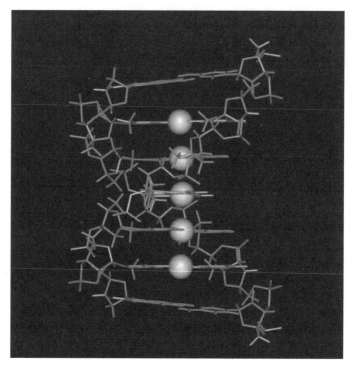

Figure 8 A plausible structure of pentanuclear Cu^{2+} complex within DNA.

V. FUTURE PROSPECTS FOR ARTIFICIAL METALLO-DNA

Such a new binding motif in DNA duplex will influence research in such diverse areas as medicinal chemistry, materials science, and bionanotechnology. Introduction of metal-induced base pairs into DNA would not only affect the assembly–disassembly processes and the structure of DNA double strands but also confer a variety of metal-based functions on DNA. Examples of one-dimensional metal arrays in solution are quite limited and, for the most part, are found only in the solid state. This strategy represents a new method for arranging metal ions in solution in a discrete and predictable fashion, leading to the possibility of metal-based molecular devices such as molecular magnets and wires.

VI. SUMMARY

DNA shows promise as a provider of a structural basis for the bottom–up fabrication of inorganic and bioorganic molecular devices. In particular, the DNA base replacement for alternative base pairing could possibly provide many versatile tools for reengineering DNA as well as biological applications. This review focuses on the most recent approaches to replace hydrogen-bonded DNA base pairs by alternative structures, including a metal-based strategy directed toward self-assembled metal arrays within DNAs.

VII. REFERENCES

1. (a) D. E. Bergstrom, in *Current Protocols in Nucleic Acid Chemistry*, S. L. Beaucage, D. E. Bergstrom, G. D. Glick, R. A. Jones, eds., Wiley, New York, 2001, Unit 1.4; (b) E. T. Kool, *Acc. Chem. Res.* **35**, 936 (2002).

2. (a) J. A. Piccirilli, T. Krauch, S. E. Moroney, S. A. Benner, *Nature* **343**, 33 (1990); (b) J. J. Voegel, S. A. Benner, *J. Am. Chem. Soc.* **116**, 6929 (1994); (c) J. D. Bain, A. R. Chamberlin, C. Y. Switzer, S. A. Benner, *Nature* **356**, 537 (1992); (d) C. Y. Switzer, S. E. Moroney, S. A. Benner, *Biochemistry* **32**, 10489 (1993); (e) J. Horlacher, M. Hottiger, V. N. Podust, U. Hübscher, S. A. Benner, *Proc. Natl. Acad. Sci. USA*. **92**, 6329 (1995).

3. (a) R. X.-F. Ren, N. C. Chaudhuri, P. L. Paris, S. Rumney IV, E. T. Kool, *J. Am. Chem. Soc.* **118**, 7671 (1996); (b) J. C. Morales, K. M. Guckian, C. Shails, E. T. Kool, *J. Org. Chem.* **63**, 9652 (1998); (c) B. M. O'Neill, J. E. Ratto. K. L. Good, D. Tahmassebi, S. A. Helquist, J. C. Morales, E. T. Kool, *J. Org. Chem.* **67**, 5869 (2002); (d) K. M. Guckian, E. T. Kool, *Angew. Chem. Int. Ed.* **36**, 2825 (1998); (e) K. M. Guckian, T. R. Krugh, E. T. Kool, *Nature Struct. Biol.* **5**, 954 (1998); (f) K. M. Guckian, T. R. Krugh, E. T. Kool, *J. Am. Chem. Soc.* **122**, 6841 (2000); (g) B. A. Schweizer, E. T. Kool, *J. Am. Chem. Soc.* **117**, 1863 (1995).

4. S. Moran, R. X.-F. Ren, E. T. Kool, *Proc. Natl. Acad. Sci. USA*. **94**, 10506 (1997).

5. (a) S. Moran, R. X.-F. Ren, S. I. Rumney, E. T. Kool, *J. Am. Chem. Soc.* **119**, 2056 (1997); (b) E. T. Kool, *Biopolymers* (*Nucl. Acid Sci.*) **48**, 3 (1998).

6. (a) D. E. Bergstrom, P. Zhang, P. H. Toma, C. A. Andrews, R. Nichols, *J. Am. Chem. Soc.* **117**, 1201 (1995); (b) P. Zhang, W. T. Johnson, D. Klewer, N. Paul, G. Hoops, V. J. Davisson, *Nucl. Acids Res.* **26**, 2208 (1998); (c) D. L. McMinn, A. K. Ogawa, Y. Wu, J. Liu, P. G. Schultz, F. E. Romesberg, *J. Am. Chem. Soc.* **121**, 11585 (1999); (d) M. Berger, A. K. Ogawa, D. L. McMinn, Y. Wu, P. G. Schultz, F. E. Romesberg, *Angew. Chem., Int. Ed.* **39**, 2940 (2000); (e) A. K. Ogawa, Y. Wu, D. L. McMinn, J. Liu, P. G. Schultz, F. E. Romesberg, *J. Am. Chem. Soc.* **122**, 3274 (2000); (f) Y. Wu, A. K. Ogawa, M. Berger, D. L. McMinn, P. G. Schultz, F. E. Romesberg, *J. Am. Chem. Soc.* **122**, 7621 (2000).

7. (a) K. Tanaka, M. Shionoya, *J. Org. Chem.* **64**, 5002 (1999); (b) M. Shionoya, K. Tanaka, *Bull. Chem. Soc. Jpn.* (*Acc.*) **73**, 1945 (2000).

8. H. Cao, K. Tanaka, M. Shionoya, *Chem. Pharm. Bull.* **48**, 1745 (2000).

9. (a) K. Tanaka, M. Tasaka, H. Cao, M. Shionoya, *Eur. J. Pharm. Sci.* **13**, 77 (2001); (b) M. Tasaka, K. Tanaka, M. Shiro, M. Shionoya, *Supramol. Chem.* **13**, 671 (2001); (c) K. Tanaka, M. Tasaka, H. Cao, M. Shionoya, *Supramol. Chem.* **14**, 255 (2002).

10. K. Tanaka, Y. Yamada, M. Shionoya, *J. Am. Chem. Soc.* **124**, 8802 (2002).

11. K. Tanaka, A. Tengeiji, T. Kato, N. Toyama, M. Shiro, M. Shionoya, *J. Am. Chem. Soc.* **124**, 12494 (2002).

12. (a) E. Meggers, P. L. Holland, W. B. Tolman, F. E. Romesberg, P. G. Schultz, *J. Am. Chem. Soc.* **122**, 10714 (2000); (b) S. Atwell, E. Meggers, G. Spraggon, P. G. Schultz, *J. Am. Chem. Soc.* **123**, 12364 (2001); (c) N. Zimmermann, E. Meggers, P. G. Schultz, *J. Am. Chem. Soc.* **124**, 13684 (2002).

13. (a) H. Weizman, Y. Tor, *Chem. Commun.* 43 (2001); (b) H. Weizman, Y. Tor, *J. Am. Chem. Soc.* **123**, 3375 (2001).

14. K. Tanaka, A. Tengeiji, T. Kato, N. Toyama, M. Shionoya, *Science* **299**, 1212 (2003).

CHAPTER 4

Organotin Macromolecules as Anticancer Drugs

Charles E. Carraher Jr.

Florida Atlantic University, Boca Raton, Florida and Florida Center for Environmental Studies, Palm Beach Gardens, Florida

Deborah Siegmann-Louda

Florida Atlantic University, Boca Raton, Florida

CONTENTS

I. GENERAL 58

II. ANTICANCER ACTIVITY OF SMALL ORGANOTIN
COMPOUNDS 59

III. MOLECULE-LEVEL STUDIES ON MONOMERIC ORGANOTIN
COMPOUNDS 62

IV. ANTICANCER ACTIVITY OF ORGANOTIN POLYMERS 65

V. FUTURE WORK 70

VI. REFERENCES 70

Macromolecules Containing Metal and Metal-Like Elements,
Volume 3: Biomedical Applications, edited by Alaa S. Abd-El-Aziz,
Charles E. Carraher Jr., Charles U. Pittman Jr., John E. Sheats, and Martel Zeldin
ISBN: 0-471-66737-4   Copyright © 2004 John Wiley & Sons, Inc.

I. GENERAL

The use of organotin-containing compounds as potential drugs in the war against cancer has occurred since the early 1970s with some but only mild success.[1-35] We more recently began to focus on the anti-cancer activity of organotin polymers that we had made for other purposes and as part of our platinum anticancer effort. Here, we will briefly review these efforts.

Hoch[1] noted that tin has a larger number of its organometallic derivatives in commercial use than any other metal. The major application (about 70%) of organotin derivatives is in PVC piping, with the organotin compounds employed as heat and light stabilizer additives. While many of these organotin compounds are monomeric, there is a move toward the use of polymeric organotin compounds because of their ability to resist ready migration. These compounds are based on oligomeric and polymeric compounds made by Carraher and coworkers.[36-41] Many other of these are employed commercially because of their biological activity. Thus, there is a reasonable literature on the general biological activities of some organotin compounds. Unfortunately, there is little information with respect to the molecular activity of these organotin compounds. Unless otherwise noted, all the alkyltin derivatives cited in this chapter are normal or linear.

Triethyltin acetate is the most toxic organotin compound to mammals with a LD_{50} rat = 4 mg/kg.[42] For comparison, the tributyltin acetate has a LD_{50} of ~400 mg/kg.[43] In general, the toxicity "cutoff" for mammals is at the butyltin, with longer chains relatively nontoxic and shorter chains somewhat toxic.[1]

Little is known about the effects of organotins in humans. What is known was reviewed in 2001.[1] Thymocytes play an important role in human immunity; 24 h after exposure in vitro to 500 ng of a variety of mono, di, and tributyltin compounds, the viability of thymocytes was halved. The natural-killer lymphocytes were affected by the presence of these butyltin compounds at a concentration of about 200 nM after 24 h.[44]

Butyltin species tend to concentrate within the liver and muscles of mammals.[45] There appears to be a lack of difference in butyltin concentrations between sexes and there is a lack of age dependence probably due to a relatively rapid excretion or blood purification.[1]

Degradation of organotin compounds in the environment occurs with progressive loss of the organic groups eventually forming the inorganic salt.

$$R_4Sn \longrightarrow R_3SnX \longrightarrow R_2SnX_2 \longrightarrow RSnX_3 \longrightarrow SnX_4$$

For polymers, degradation will probably occur not only through a variety of routes including the loss of alkyl (aryl) groups but also through both purely physical and enzymatic assisted hydrolysis, with reactions releasing the various organotin and non-organotin moieties. In Nature, a wide variety of bacteria are found to degrade organotin compounds, including *P. aeruginosa*, *P. putida C*, *A. faecalis*, and *C. vulgaris*.[1] Organotin compounds are also known to undergo degradation through both nucleophilic and electrophilic attack.

It has been reported that trialkyltin compounds such as tributyltin and triphenyltin are metabolized sequentially to dialkyltin, monoalkyltin, and inorganic tin compounds.[46,47] Other studies focusing on the actual active form of the organotin indicate that only the original forms of the organotin compounds are found in the cancer cells, consistent, and that the active form of the tin is the original compound and not a metabolite.[48]

II. ANTICANCER ACTIVITY OF SMALL ORGANOTIN COMPOUNDS

While the first organotin compound was tested for antitumor activity in 1929,[21] no systematic study was undertaken and only recently (as of 2003) has the foundation for using organotin compounds as anticancer agents been explored. Since the discovery by Furst[49] that metal chelation plays a role in the cure and cause of malignancy, effort has focused on the use of real and potential chelating agents. Many of these have included metal sites as an essential part of the chelating agent.[49]

Here we briefly review some of the evaluations of organotin compounds with respect to their ability to inhibit cancer. Most of these studies involve simple cell culture studies, but they may be instructive as to what constitutes important structural considerations. The anticancer activity of small organotin compounds is well established. Here we will look at some specific examples to lay the foundation for our efforts.

The ability to combine with some biologically important material is important for organotin compounds to inhibit cancer cell growth. This combination can be through simple association or complex formation utilizing the unfilled *d* orbitals on tin. The combinations can also occur through release of groups chemically bound to the tin, such as the halides, which then free the organotin moiety to chelate with the biologically important material.

We have found that some organometallic dichlorides exhibit antitumor activity but believe that this activity is due in part to the hydrolysis and subsequent formation of hydrochloric acid. The toxicity is a result of the cells' primary or sole response to the hydrochloric acid. Even so, a number of studies involve the use of organotin halides, and it is possible that the organotin halides are themselves active agents. Most studies involving organotin halides utilize the chloride since this halide was generally found to be best in early studies.[21]

Atsushi and co-workers in 1973 reported the high affinity of tin for tumors.[50] This was later used when tin-labeled technetium complexes were employed as imaging agents to locate tumors.

In the evaluation of a number of studies, Saxena and Huber [21] found that the most active organotin species were the dialkyltin compounds. Brown[52] found that triphenyltin acetate exhibited antitumor activity in mice, while triphenyltin chloride was inactive. She believed that the degree of water solubility was an important factor in the anticancer activity of the organotin compounds.

Ozaki and Sakai[53] patented 2',3'-*O*-dialkylstannyl derivatives of 5-fluorouri-dine as anticancer agents (**1**). The compounds caused the shrinkage of solid tumors when injected directly. This early study shows the attempts to couple known antitumor drugs with organotin compounds.

In 1980 Crowe et al[54] published the first report on the anticancer activity of a series of diorganotin dihalides and complexed products of the general formula R_2SnX_22L, where 2L (bidentate) = phenanthroline, bipyridyl, 2-aminomethylpyri-dine; L (monodentate) = dimethylsulfoxide, pyridine; X = halides; and R = methyl, ethyl, propyl, butyl, or phenyl. They postulated that a somewhat stable organotin complex is needed for good activity. The diethyltin dichloride compounds were found to be the most active.

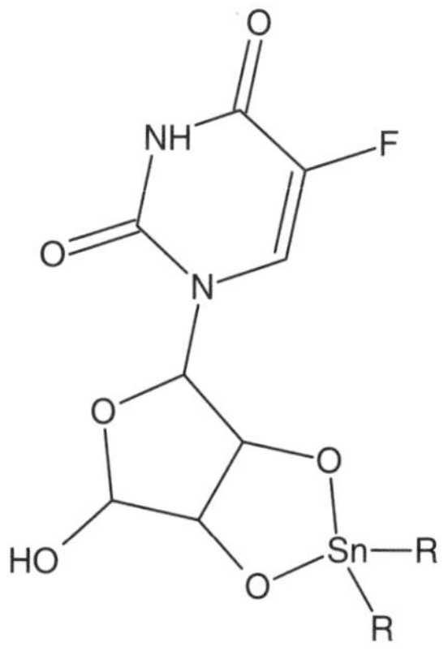

1

Other early studies by Crowe and coworkers also showed that diethyltin and diphenyltin complexes generally showed the highest activity. Further, ligands that contain strongly complexing nitrogen ligands also showed good activity. It must be remembered in these early studies that the complexes were probably a mixture of structures with some having octahedral structures where the nitrogen atoms formed complexes through vacant *d* orbitals on the organotin. These structures are not the same as those where groups are substituted for the halides, so results from the study of these complexes must be viewed with this in mind.

The activity of *cis*-DDP is believed to be the result of selective binding of the *cis*-DDP with DNA bases that act to effectively inhibit further DNA synthesis and thus subsequent curtailment of the cancer. A number of monomeric organotin drugs were made of the general form $R_2SnX_2L_2$, where R was methyl, ethyl, or similar; (X = halide and SCN; L = oxygen or nitrogen donor ligand). Some of these organos-tannane drug compounds showed good activity against several cancers, including P338 lymphocytic leukemia, on the same level as *cis*-DDP, but showed little or no nephrotoxicity that is associated with *cis*-DDP. It was believed that because the X−M−X bond angles of the platinum and tin are similar, a similar biological mechanism might occur.[55] However, no correlation was found between activity and bond angle. From this it was concluded that the mechanism for activity was probably dis-similar to that of *cis*-DDP.[55]

Cardarelli and coworkers[56,57] found that soluble organotin compounds are con-centrated in the thymus gland. They believed that the organotin in the thymus is then processed into one or more biochemicals that act as anticarcinogens or antioncogens. They isolated a tin-bearing steroid compound, **2**. On this basis, Cardarelli and coworkers patented several organotin compounds of steroids that exhibited good anti-tumor activity. However, no structure–activity relationships were found.

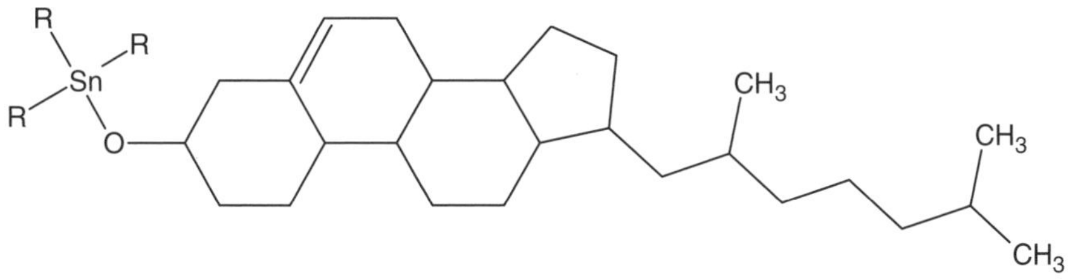

2 (organotin steroid)

Meinema et al.[58] screened a number of complexes of the general formula RR′Sn(CH$_2$COOMe)$_2$ and RR′SnO. Along with the usual cancer screening, they found that the active organotin compounds induced filamentous growth in bacteria indicative of interaction with DNA.

Kovala-Demertzi et al. used the complexing ability effect of 1-methyl-imida-zoline-2(3*H*)-thione (Hmimt) and imidazoline-2(1,3*H*)-thione (Himt) with several organotin compounds.[26] Hmimt is used in the treatment of thyroid disorders.[59]

In these products the Hmimt and Himt were connected to the tin through the sulfur. One general Hmimt structure for these compounds is given as **3**. In general, the in vitro antitumor activity of these compounds with sulfur-donating heterocyclic ligands was generally poor, significantly lower than similar compounds with Sn–O linkages.[60,61]

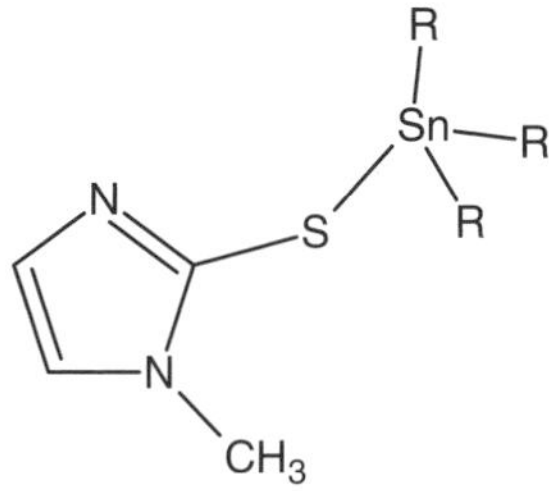

3

This is consistent with the importance of the Sn linkage in determining the overall activity.

Gielen and Willem[12] studied the in vitro cytotoxicity of diorganotin trimethoxy benzoates against 60 human tumor cell lines. In general, they found that the dibutyltin compounds were more active against human cancer cell lines than were the diethyltin derivatives.

While most studies found that the dibutyltin derivatives are most active, some have reported other prescreening results. For instance, it was found that diethyltin and diphenyltin derivatives were more active during in vivo tests against P388 and L1210 murine leukemias in comparison to the dibutyltin compounds.[62]

Some compounds were found to give lower toxicities in vivo against certain cancers in comparison to cisplatin, the most widely used anticancer drug.[63,64]

But, there are numerous reports detailing superior activities for the organotin, generally the dibutyltin derivative, in comparison to cisplatin.[10,30,66,67]

The LD_{50} for dibutyltin acetate is ~400 mg/kg. The LD_{50} for cisplatin is 12 mg/kg. By comparison, dibutyltin compounds appear to be less toxic than cisplatin.[68]

The evaluation of organotin compounds with respect to their ability to inhibit cancer cell growth continues.

III. MOLECULE-LEVEL STUDIES ON MONOMERIC ORGANOTIN COMPOUNDS

Following are several studies indicating that the activity of organotin compounds on cancer cells is varied and complex. Currently, it is uncertain whether there is in fact a common target(s) for organotins.

Pellerito and co-workers have coupled a number of known drugs including amoxicillin, penicillin G, chloramphenicol, cycloserine, and methicillin, with organotins.[14–20] A general structure for the methicillin compound is given in **4**.

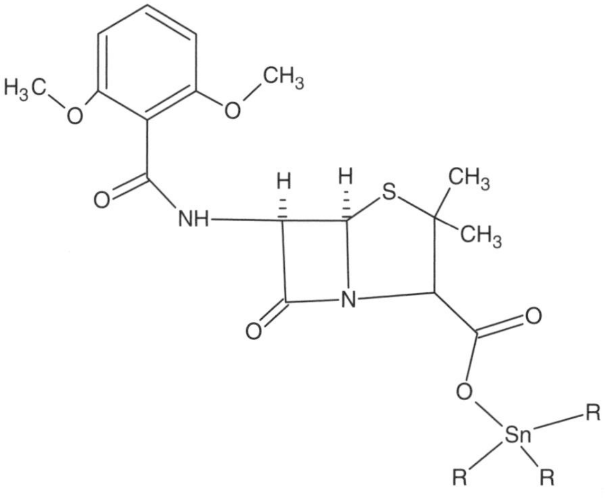

4

These compounds exhibited good inhibition of a number of cancer cell lines. To help determine the target site(s), this group focused on studying the mitotic chromosomes of the fish cypriniform *Rutilus rubilio* with respect to the activity of the organotin compounds. They found that fish treated with diorganotin derivatives showed differentially stained chromosome areas with the granular zones deeply stained along the chromosomal body. They also found arm breakages and sidearm bridges (pseudochiasmata) in the chromosomes. This is consistent with the organotin compounds causing chromosome anomalies. The most frequently observed chromosome anomaly is the presence of gaps or achromatic lesions. The anomaly may be linked to a different degree of DNA condensation. The methyl derivatives exert a lower cytotoxicity than butyl and phenyl organotin derivatives.

Yamabe et al.[48] studied the enhancement of androgen-dependent transcription and cell proliferation by tributyltin, triphyenyltin, and other organotin compounds. As an aside, the total chemical nature of the organotin compounds was not cited in the paper and illustrates the gulf that can exist between the chemist and the medical researcher. Even so, their results showed that the organotin compounds enhanced the DNA synthesis and expression of endogenous androgen receptor, AR, target genes such as prostate specific antigen, but not the expression of AR itself. Further, trialkyltin compounds can activate AR-mediated transcription in mammalian cells through a target site other than the ligand-binding site of AR.

In vitro and in vivo studies have indicated that different organotin compounds show different cytotoxic effects on different types of cell lines. Thus, tributyltin and triphenyltin showed immunosuppressive behavior while trimethyltin and triethyltin compounds were neurotoxic.[69] It is known that different organotin compounds show different effects on Na^+/K^+-ATPase activity.[70] Further, different organotin halides have shown different effects on voltage-gated potassium currents in lymphocytes and neuroblastoma cells.[71]

Inhibition of macromolecular synthesis and of the mitochondrial energy metabolism, reduction of DNA synthesis, and direct interaction with the cell membrane have all been implicated in the organotin-induced cytotoxicity. Interaction of the tributyltin compounds with the cell membrane was reported to cause the ion channels to open. This is followed by an influx of extracellular Ca^{2+} ion giving an increased cytosolic calcium concentration.[72] This extracellular calcium ion influx has been indicated in tributyltin-induced apoptosis in mouse thymocytes.[73]

Tributyltin moieties are also believed to bind to a part of the ATPsynthase complex, resulting in inhibition of mitochondrial ATP synthesis and distortion of the proton gradient.[74] The result of the above is that electrons are diverted from the respiratory chain that forms the reactive oxygen species (ROS). Both the decrease in ROS and increased concentration of intracellular calcium ion are believed to be the major factors involved in triorganotin-induced apoptosis in many cell lines.

Ray et al.[75] studied the different cytotoxic effects for a number compounds, including various tributyltin halobenzoates on human leukemic K562 cells. They also studied the role of extracellular calcium ion influx and ROS formation in inducing apotosis in this cell line.[76] Human breast cancer MCF-7 cells were also employed in the studies.[77] The K562 has a low membrane cholesterol content, while the MCF-7 has an excess membrane cholesterol content.[75,77] Membrane cholesterol content is believed to be responsible for the fluidity of the membrane that allows the calcium ion movement across the membrane to be altered.[78]

The tributyltin benzoates were toxic to the K562 cells.[75] Tributyltin benzoate (TBSB)-treated K562 cells showed a greater initial extracellular calcium ion influx compared with the tributyltin halobenzoates. This is consistent with an early activation of endonucleases and DNA fragmentation in the TBSB-treated cells. This is also consistent with an earlier report by Chow et al.[73] where the cytosolic-free calcium ion concentration increases because of an extracellular calcium ion influx, an inhibition of the calcium ion extrusion system, and the release of calcium from the intracellular reserves. These, and other studies, suggest involvement of nonspecific cation channels in the extracellular calcium ion influx that leads to an increase in observed cell viability. When the calcium channels were blocked with either verapamil or nefidipine, the extracellular calcium ion influx was partly inhibited and apotosis observed.

K562 cells were found to undergo apoptosis at a lower concentration of the tributyltin-2,6-difluorobenzoate and after a shorter treatment than did MCF-7 cells. Again, these and other results were consistent with the involvement of both the influx of extracellular calcium ion and ROS in the induction of apoptosis in the K562 cells. Because of the low cholesterol content in the K562 cells, the tributyltin compounds were able to fluidize the membrane leading to an opening of the ion channels, allowing an influx of the extracellular calcium ions.By comparison, the excess cholesterol content in the MCF-7 cells may mean that the organotin compounds are not able to fluidize the membrane and therefore block the extracellular calcium ion influx.

Barbieri and coworkers[79,80] have studied the antiproliferative activity and interactions of cell-cycle-related proteins utilizing the organotin compound containing a quinolizidine derivative. The particular compound studied was triethyltin lupinylsulfide hydrochloride (**5**). In early tests this compound was found to show good cancer cell line inhibition, so more extensive studies were undertaken.

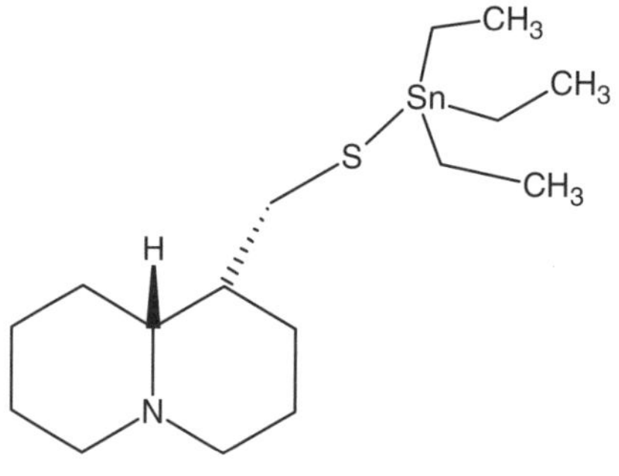

5 (triethhyltin lupinylsulfide)

The cytocidal effects shown by this compound were more consistent with necrosis or delayed cell death rather than apoptosis as shown by morphologic observations, DNA fragmentatin analysis, and flow cytometry.[81] A number of genes that play an important role in the G1/S phase transition were eliminated as primary targets of the compound with the results indicative of a direct effect of the compound on macromolecular synthesis and cellular homeostasis. This result is consistent with studies of other organotin compounds where DNA does not appear to be the primary target. Instead, perturbation of homeostasis, impairment of mitochondrial functions, and inhibition of protein synthesis may be the main reasons for the anticancer response to organotin compounds.[82,83]

Stridh et al.[84] reported that organotin compounds such as triphenyltin and tributyltin kill target cells by triggering apoptosis by caspase activation at low drug concentrations, but bring about necrotic cell death at higher organotin concentrations. This suggests that apotosis and necrosis may not be totally distinct pathways but that there is some overlap.[81]

The variation in inhibition mechanism from that of cisplatin is important since it means that organotin compounds offer an additional and different role than cisplatin in combating cancer.

Such studies allow a better understanding of the possible mechanisms involved in organotin anticancer activity but also show that such mechanisms may vary with the particular cell lines as well as with the nature of the specific organotin compound. The second finding indicates that particular organotin compounds should be studied on a variety of cell lines before being cast aside as being inactive.

IV. ANTICANCER ACTIVITY OF ORGANOTIN POLYMERS

At least two approaches are taken in the design of metal-containing polymeric drugs. The first is to design the macromolecule such that it is the vehicle that delivers the active component, the metal-containing moiety. The second is to design the macromolecule as the actual drug. In our efforts we have considered the organotin polymers where the polymer may act as a drug itself, and/or as a controlled release agent. We have described elsewhere the possible advantages to having the metal-containing active component as part of the polymer drug.[85]

Briefly, inclusion of the active ogranotin moiety as part of a polymer drug should

1. Limit movement of the biologically active drug. Because of their size, polymers are not as apt to easily pass through membranes present in the body Polymers with chain lengths of about 100 units and greater typically are unable to move easily through biological membranes. Restricted movement may prevent buildups in the kidneys and other organs, thereby decreasing renal and other organ damage along with associated effects. Further, the active organotin moiety may be released slowly, reducing the exposure of organs to large concentrations of metal-containing moiety. This should also increase the concentration of organotin in the beneficial form in the body, permitting lower effective doses of the drug to be used.

2. Enhance activity through an increased opportunity for multiple bonding at a given site.
3. Increase delivery of the bioactive moiety and decrease toxicity. It is believed that the polymeric nature of the drug will "protect" the active portion through steric constraints restricting the approach of water to the active site. Also contributing to this protection is the fact that the polymer is not as hydrophillic as the organotin moiety itself as shown by the lack of water solubility of the organotin-containing polymers. Aqueous solubility is inhibited by some organotin-containing small compounds.
4. Possibly bypass the cell's defense system that is armed because of the presence of the invasion of other chemotherapeutic (chemo) drugs. Studies are indicating that introduction of chemo drugs into cells causes the buildup of "housekeeping" proteins that are rather general in their ability to select and remove foreign compounds present in the cell. This may be a principal reason why chemo treatments lead to resistance to drugs, even to drugs that have not been previously used. It is possible that the polymeric nature of the organotin polymers will discourage the housekeeping proteins from removing these polymeric drugs thus allowing the polymers to function as anticancer drugs under conditions where smaller organotin-containing drugs are not successful.

Initially we screen materials for biological effects utilizing normal Balb/3T3 cells because they are easier to handle. Active compounds are then tested against cancer cell lines.

Our initial study involved products derived from ampicillin.[86–89] The structure shown below (**6**), is for the N-Sn-O product. Because ampicillin is not symmetric, two additional repeat units are possible designated as O-Sn-O and N-Sn-N.

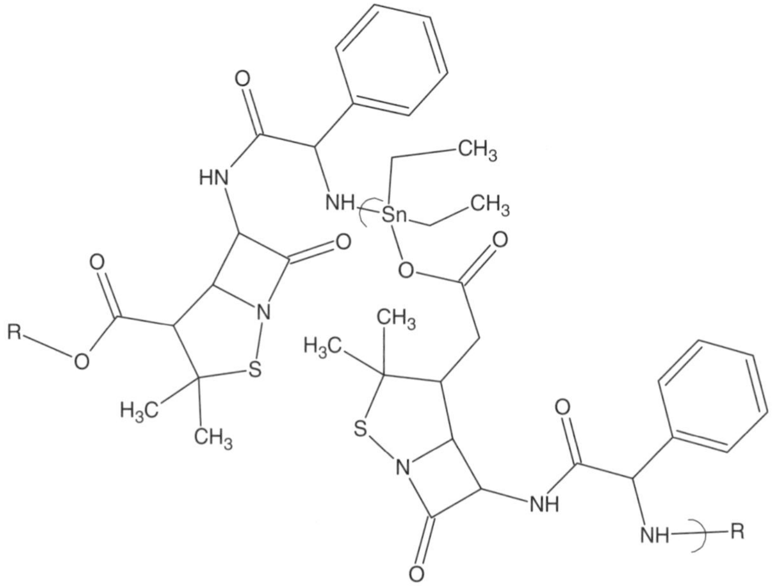

6

This study focused on the diethyltin derivative since this was most active against bacteria. The polymer showed equal or better ability to inhibit cell growth in comparison to diethyltin dichloride with decent inhibition at 5 µg/mL.

The diethyltin polymers derived from glycylphenylalanine and 4-aminobenzoic acid were examined. The glycylphenylalanine polymer showed better than 50% inhibition at 10 µg/mL while the polymer from 4-aminobenzoic acid showed about 80% inhibition at 5 µg/mL.

Then two cancer cell lines were studied.[87-89] The HTB 75 cell line is from a cancer patient with ovarian cancer that had previously been treated with cytoxan, adriamycin, 5-fluorouracil, and Fur IV. The HTB161 cell line is from a cancer patient with ovarian cancer that had been treated with adriamycin, cyclophosphamide, and cisplatin. This cell line is known to be drug-resistant. The polymers from ampillicin, 4-aminobenzoic acid and glycylphenylalanine all showed mild inhibition of the cell lines at 50 µg/mL.

Then the diethyltin and dimethyltin polymers resulting from reaction with cephalexin were tested.[90] The repeat unit structure is given as **7** for the dimethyltin product.

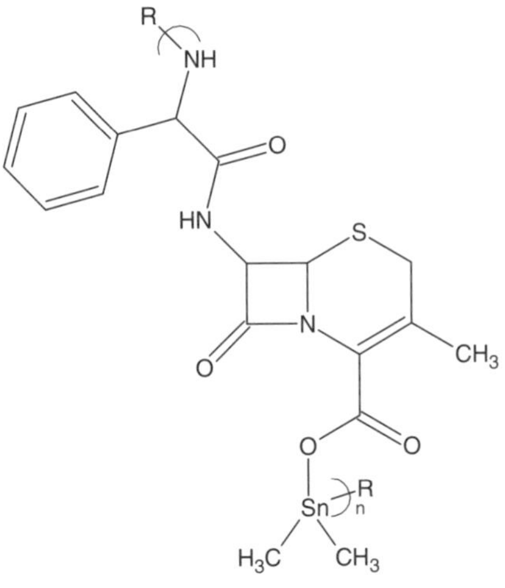

7

The diethyltin and also the dimethyltin products were studied because of their outstanding antibacterial activities. Both showed good inhibition of the Balb/3T3 cells at about a polymer concentration of 10 µg/mL. Again, the polymers showed mild activity against the HTB 75 cancer cell line. Greater than 50% inhibition was found for the diethyltin polymer at 30 µg/mL.

Next, the polymer derived from the plant growth hormone kinetin was studied.[91] A representative structure is given in **8** for the product from dimethyltin dichloride.

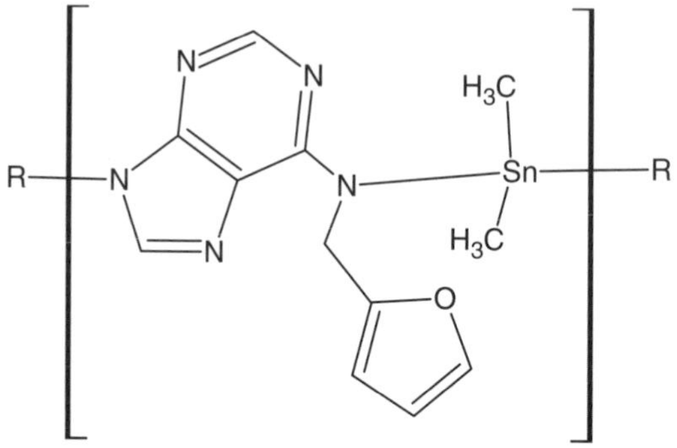

8

Inhibition of Balb/3T3 cells was studied as a function of the alkyltin. Kinetin, itself, does not inhibit cell growth. The general trend is that activity increases as the size of the alkyl group increases up to the butyl group, after which it again decreases. To test the lower level of activity for the most active polymer, continued dilutions of the dibutyltin-kinetin polymer solution were made. Good activity, >50% inhibition, was found at about 0.20 mg/mL. By comparison, *cis*-DDP itself shows good activity in the 0.10 μg/mL range with about a 50% inhibition of many cells. Thus the organotin polymers act in about the same concentration range as *cis*-DDP.

The good activity of the kinetin-containing polymers is not unexpected since kinetin itself is a substituted aminopurine. Further, kinetin is known to directly interact with at least plant nucleic acids.

A similar study was completed except for norfloxacin-containing polymers. The general repeat unit is given in **9** for these products.[92]

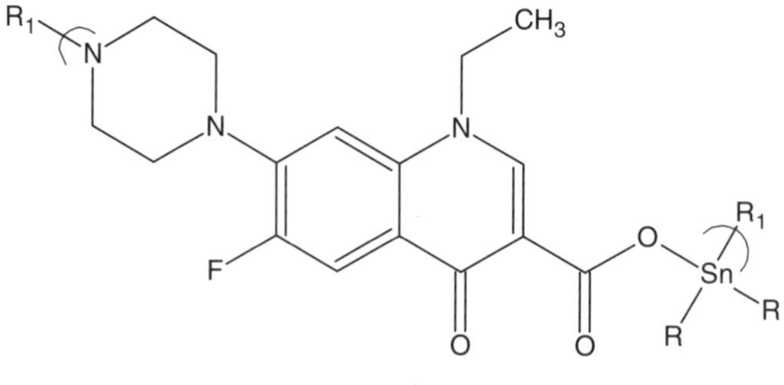

9

Again, the order of activity increased to the butyl group after which it decreased. The dipropyltin derivative showed good activity to 0.40 μg/mL while the dibutyltin showed good activity to 0.10 μg/mL.

While we have coupled the organotin portion with known drugs, we have found that even very simple structures give good inhibition of cell growth. For instance, the diethyltin /adipic acid polymer, the general structure of which is given in **10**, shows good inhibition at 25 µg/mL.[93]

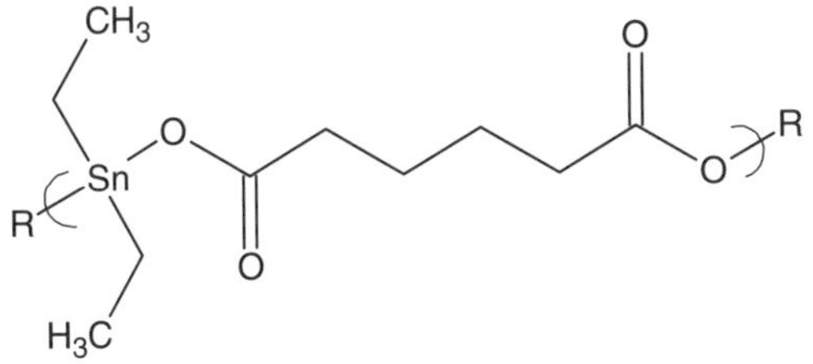

10

Further, the presence of organotin does not guarantee activity against cell growth. The analogous product to **10** except derived from the phenylmethyltin dichloride exhibits no activity at the 25 µg/mL level.[94]

We also tested a number of diol containing organotin polymers, focusing on dibutyltin and diphenyltin derivatives.[94] The structure for the 1,4-butanediol product with dibutyltin dichloride is given in **11**. This product inhibited cell growth at 0.25 µg/mL. The analogous diphenyltin product inhibits growth at 1.0 µg/mL.

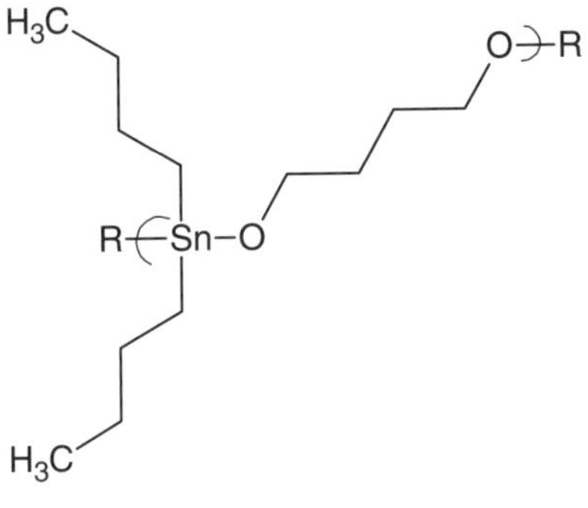

11

For comparison, the 1,6-hexanediol products inhibited growth at 5 µg/mL for the dibutyltin product and at 2 µg/mL for the diphenyltin product. Thus, the nature of the connecting linkage is important in determining the extent of inhibition.

Small differences in the nature of the diol appear to be very important in determining the activity of the polymer. Thus, the diphenyltin derivative of 1,4-butenediol was tested and showed good activity at 1.0 µg/mL while the dibutyltin derivative, **12**, showed good activity to 0.025 µg/mL. The butenediol can be cis or trans.

The proportion of cis and trans for this product is not known, but the material is sold as a mixture. Most of it is the cis derivative as shown in **12**.

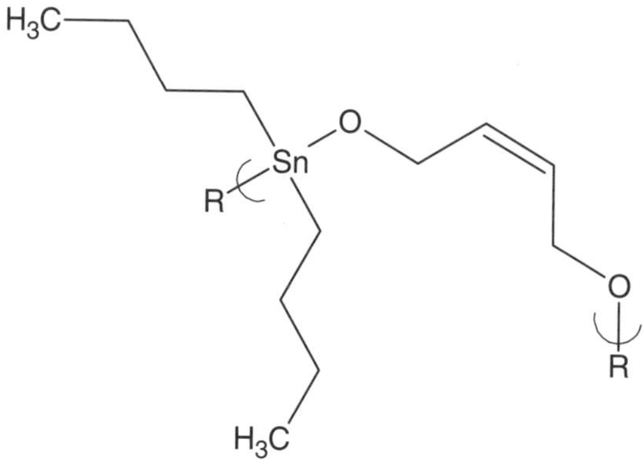

12

V. FUTURE WORK

Molecule level studies aimed at understanding the mechanism(s) of activity of organotin in the inhibition of cancer cells need to be continued and extended to the polymeric derivatives. While we have found that the activity appears to peak for the dibutyltin derivatives, the dipentyltin and dihexyltin derivatives need to be studied to see if, in fact, the dibutyltin products are the most active. The importance of the nature of the nonalkyl connective linkages needs to be determined. Thus, a wider array of products needs to be evaluated. Live-animal tests need to be conducted on the most promising products. Finally, any synergistic effects that may result from the combined application of platinum and tin compounds should be studied.

VI. REFERENCES

1. M. Hoch, *Appl. Geochem.* **16**, 719 (2001).

2. F. Barbieri, M. Viale, F. Sparatore, G. Schettini, A. Favre, C. Bruzzo, F. Novelli, A. Alama, *Anti-Cancer Drugs* **13**, 599 (2002).

3. M. Gielen, R. Willem, A. Bouhdid, D. Vos, C. Kuiper, G. Veerman, G. Peters, *In Vivo* **9**, 59 (1995).

4. M. Gielen, *Metal-Based Drugs* **2**, 99 (1995).

5. M. Gielen, F. Kayser, O. Zhidkova, V. Kampel, V. Bregadze, D. Vos, B. Mahieu, R. Willem, *Metal-Based Drugs* **2**, 37 (1995).

6. P. Carpinelli, S. Bartolucci, F. Ruffo, *Anti-Cancer Drug, Design* **10**, 43 (1995).

7. Y. Arakawa, *Main Group Metal Chem.* **17**, 225 (1994).

8. J. Peng, H. Su, *Yaoxue Xuebao* **29**, 406 (1994).

9. Y. Arakawa, *Bio. Chem. Trace Elem.* **4,** 129 (1993).

10. M. Gielen, R. Willem, *Anticancer Res.* **12,** 1323 (1992).

11. J. Xiao, J. Cui, Y. Su, J. He, Y. Junen, *J. Chin. Pharm. Sci.* **2,** 45 (91993).

12. M. Gielen, D. deVos, A. Meriem, M. Boualam, A. El Mkloufi, R. Willem, *In Vivo* **7,** 171 (1993).

13. F. Capolongo, A. Giuliani, M. Giomini, R. U. Russo, *J. Inorg. Biochem.* **49,** 275 (1993).

14. L. Pellerito, F. Maggio, M. Consiglio, A. Pellerito, G. Stocco, S. Grimaudo, *Appl. Organomet. Chem.* **9,** 227 (1995).

15. F. Maggio, A. Pellerito, L. Pellerito, S. Grimaudo, C. Mansueto, R. Vitturi, *Appl. Organomet. Chem.* **8,** 71 (1994).

16. R. Vitturi, C. Mansueto, A. Gianguzza, F. Maggio, A. Pellerito, L. Pellerito, *Appl. Organomet. Chem.* **8,** 509 (1994).

17. H. Baratne Jankovics, L. Nagy, F. Longo, T. Fiore, L. Pellerito, *Magyar Kemiai Foly.* **107,** 392 (2001).

18. R. Vitturi, B. Zava, M. Colomba, A. Pellerito, F. Maggio, L. Pellerito, *Appl. Organomet. Chem.* **9,** 561 (1995).

19. L. Pellerito, F. Maggio, T. Fiore, A. Pellerito, *Appl. Organomet. Chem.* **10,** 393 (1996).

20. A. Pellerito, T. Fiore, C. Pellerito, A. Fontana, R. Di Stefano, L. Pellerito, M. Cambria, C. Mansueto, *J. Inorg. Biochem.* **72,** 115 (1998).

21. A. K. Saxena, F. Huber, *Coord. Chem. Revs.* **95,** 109 (1989).

22. E. Bulten, H. Budding, U.S. Patent 4,547,320 (Oct. 15, 1985) to Nederlandse Centrale Organisatie Voor.

23. F. Caruso, *J. Med. Chem.* **36,** 1168 (1993).

24. M. Boualam, M. Gielen, A. El Khloufi, D. deVos, W. Rudolph, U.S. Patent 5,382,597, (Jan. 17, 1995) to Pharmachemie B.V.

25. M. Gielen, R. Willem, A. Bouhdid, D. de Vos, U.S. Patent 5,559,147, (Sept. 24, 1996) to Pharmachemie B.V.

26. D. Kovala-Demertzi, P. Tauridou, U. Russo, M. Gielen, *Inorg. Chem. Acta* **239,** 177 (1995).

27. B. Biddle, J. Gray, *Appl. Organomet. Chem.* **5,** 439 (1991).

28. S. Hiu, M. Huang, D. Shi, T. Huang, J. Wan, Z. Huang, *Jiegou Huaxue* **9,** 264 (1990).

29. T. Fujimura, T. Sato, S. Hayasaka, S. Furusawa, H. Kawauchi, S. Kenichi, Y. Takayanagi, *Annu. Rept. Tohoku Coll. Pharm.* **36,** 239 (1989).

30. V. Narayanan, M. Nasr, K. Paull, *NATO ASI Series, Series H, Cell Biol.* **37,** 201 (1990).

31. A. Penninks, M. Bol-Schoenmakers, W. Seinen, *NATO ASI Series, Series H, Cell Biol.* **37,** 169 (1990).

32. A. Saxena, F. Huber, *Coord. Chem. Rev.* **95,** 109 (1989)

33. A. Penninks, P. Punt, M. Bol-Schoenmakers, H. Van Rooijen, W. Seinen, *Silicon, Germanium Lead Compds.* **9,** 367 (1986).

34. L. Sherman, *Silicon, Germanium, Tin, Lead Compds.* **9,** 323 (1986); L. Sherman, F. Huber, *Appl. Organomet. Chem.* **2,** 65 (1988).

35. N. Brown, *Diss. Abst. Int. B* **33,** 5356 (1986).

36. R. Wei, L. Ya, W. Jinguo, X. Qifeng, in *Polymer Materials Encyclopedia*, J. Salamone, ed., CRC Press, Boca Raton, FL, 1996, p 4826.

37. C. Carraher, *Angew. Makromol. Chemie.* **31,** 115 (1973).

38. C. Carraher, R. Dammeier, *Polym. Preprints* **11**(2), 606 (1970).

39. C. Carraher, R. Dammeier, *J. Polym. Sci., A-1* **8,** 3367 (1970).

40. C. Carraher, R. Dammeier, *Makromol. Chemie* **135,** 107 (1970).

41. C. Carraher, R. Dammeier, *J. Polym. Sci., A-1* **10,** 413 (1972).

42. P. J. Smith, J. Luitjen, O. Klimmer, *Toxicological Data on Organotin Compounds*, ITIR Publicaton 538, International Tin Research Institute, London, 1978.

43. E. Bulten, H. Meinema, in *Metals and Their Compounds in the Environment*, E. Merian, ed., VCH, Weinheim, 1991.

44. K. Kannan, K. Senthilkumar, J. Giesy, *Env. Sci. Technol.* **33**, 1776 (1999).

45. S. Tanabe, M. Prudente, T. Mizuno, J. Hasegawa, H. Iwata, N. Miyazaki, *Env. Sci. Technol.* **32**, 193 (1998).

46. A. Kanetoshi, *Eisei Kagaju*, **29**, 303 (1983).

47. T. Horiguchi, H. Shiraishi, M. Shimizu, M. Morita, *Env. Pollut.* **95**, 85 (1997).

48. Y. Yamabe, A. Hoshino, N. Imura, T. Suzuki, S. Himeno, *Toxicol. Appl. Pharmacol.* **169**, 177 (2000).

49. A. Furst, *Chemistry of Chelation in Cancer*, Thomas Springfield, New York, 1963.

50. A. Atsushi, K. Hisada, I. Ando, *Radioisotopes* **22**, 7 (1973).

51. M. Yamaguchi, K. Sugii, S. Okada, *J. Toxicol. Sci. Jpn.* **5**, 238 (1981).

52. N. M. Brown, Ph.D. thesis, Clemson Univ., Clemson, SC, 1981.

53. S. Ozaki, H. Sakai, Jpn Patent 76,100,089 (1975).

54. A. J. Crowe, P. Smith, G. Atassi, *Chem.-Biol. Interact.* **32**, 171 (1980).

55. I. Omae, *Organotin Chemistry*, Elsevier, New York, 1989.

56. N. Cardarelli, B. Cardarelli, E. Libby, E. Dobbins, *Aust. J. Exp. Biol. Med. Sci.* **62**, 209 (1984).

57. N. Cardarelli, S. Kanakkant, U.S. Patent 4541956, 1984.

58. H. Meinema, A. Liebregts, H. Budding, E. Bulten, *Rev. Silicon, Germanium, Tin Lead Compds.* **2–3**, 157 (1985).

59. A. Taurog, *Endocrinology*, **98**, 1031 (1976).

60. P. Tauridou, U. Russok D. Marton, G. Valle, D. Kovala-Demertzi, *Inogr. Chim. Acta* **231**, 139 (1995).

61. M. Boualam, J. Meunier-Piret, M. Biesemans, R. Willem, M. Gielen, *Inorg. Chim. Acta* **198–200**, 249 (1992).

62. A. J. Crowe, in *Metal-Based Antitumor Drugs*, Vol. 1, M. Gielen, ed., Freund, London, 1988, p. 103.

63. H. Meinema, A. Liebregts, H. Budding, E. Bulten, *Revs. Silicon, Germanium, Tin, Lead Compds.* **8**, 157 (1985).

64. G. Atassi, *Revs. Silicon, Germanium, Tin Lead Compds.* **8**, 219 (1985).

65. M. Gielen, J. Meunier-Piret, M. Biesemans, R. Willem, *Appl. Organomet. Chem.* **6**, 59 (1992).

66. M. Boualani, M. Gielen, A. Meriem, D. de Vos, R. Willem, Eur. Patent 902023167 (Dec. 1990).

67. M. Ross, M. Gielen, P. Lelieveld, D. de Vos, R. Willem, *Anticancer Res.* **11**, 1089 (1991).

68. *The Toxic Substances List*, 1974 ed. U.S. Dept. Health, Education, and Welfare, National Institute for Occupational Safety and Health, Rockville, MA, 1974.

69. B. Viviani, A. Rossi, S. Chow, P. Nicotera, *Neurotoxicology* **16**, 19 (1995).

70. P. Samuel, S. Roy, K. Jaiswal, J. Rao, *J. Appl. Toxicol.* **18**, 383 (1998).

71. M. Oortgiesen, E. Visser, H. Vijverberg, W. Seinen, *Naunyn-Schmiedebergs Arch. Pharmacol.* **353**, 136 (1996).

72. T. Aw, P. Nicotera, L. Manzo, S. Orrenius, *Arch. Biochem. Biophys.* **283**, 46 (1990).

73. S. Chow, G. Kass, N. McCabe, S. Orrenius, *Arch. Biochem. Biophys.* **298**, 143 (1992).

74. E. Corsini, B. Viviani, M. Marinovich, C. Galli, *Toxicol. Appl. Pharmacol.* **145**, 74 (1997).

75. D. Ray, K. Sarma, A. Antony, *Life* **49**, 519 (2000).

76. M. Maccarone, W. Nieuwenhuizen, W. Dulles, H. Catani, M. Melino, C. Veldink, J. Vliegenthart, A. Agro, *Eur. J. Biochem.* **241**, 297 (1996).

77. B. Cyproani, C. Tabacik, B. Descomps, *Biochem. Biophys. Acta* **972**, 167 (1988).

78. M. Maccarrone, L. Bellincampi, G. Melino, A. Agro, *Eur. J. Biochem.* **253,** 107 (1998).

79. M. Cagnoli, A. Alama, F. Barbieri, F. Novelli, C. Bruzzo, F. Sparatore, *Anti-Cancer Drugs* **9,** 603 (1998).

80. F. Barbieri, M. Viale, F. Sparatore, A. Favre, M. Cagnoli, C. Bruzzo, F. Novelli, A. Alma, *Anticancer Res.* **20,** 977 (2000).

81. F. Barbieri, F. Sparatore, M. Cagnoli, C. Bruzzo, F. Novelli, A. Alama, *Chemico-Biol. Interact.* **134,** 27 (2001).

82. A. H. Penninks, in *Tin-Based Antitumor Drugs*, M. Gielen, ed., Springer, Berlin, 1990, pp. 169–190.

83. M. Aschner, J. Aschner, *Neurosci. Biobehav. Rev.* **16,** 427 (1992).

84. H. Stridh, S. Orrenius, M. Hampton, *Toxicol. Appl. Pharmacol.* **156,** 141 (1999).

85. D. Sigmann-Louda, C. Carraher, *Macromolecules Containing Metals and Metalloids*, Vol. 3, Wiley, 2004.

86. C. Carraher, F. Li, D. Siegmann-Louda, C. Butyle, J. Ross, *Polym. Mater. Sci. Eng.* **77,** 499 (1997).

87. C. Carraher, F. Li, D. Siegmann-Louda, C. Butler, S. Harless, F. Pflueger, *Polym. Mater. Sci. Eng.* **80,** 363 (1999).

88. D. Siegmann-Louda, C. Carraher, F. Pfueger, D. Nagy, *Polym. Mater. Sci. Eng.* **84,** 658 (2001).

89. D. Siegmann-Louda, C. Carraher, J. Ross, F. Li, K. Mannke, S. Harless, *Polym. Mater. Sci. Eng.* **81,** 151 (1999).

90 D. Siegmann-Louda, C. Carraher, F. Pflueger, J. Coleman, S. Harless, H. Leuing, *PMSE,* **82,** 83 (2000).

91. D. Siegmann-Louda, C. Carraher, D. Chamely, A. Cardoso, D. Snedden, *Polym. Mater. Sci. Eng.* **86,** 293 (2002).

92. D. Siegmann-Louda, C. Carraher, M. Graham, R. Doucette, L. Lanz, *Polym. Mater. Sci. Eng.* **87,** 247 (2002).

93. D. Siegmann-Louda, C. Carraher, F. Pflueger, D. Nagy, J. Ross, *Functional Condensation Polymers*, Kluwer, New York, 2002.

94. D. Siegmann-Louda, C. Carraher, unpublished results.

CHAPTER 5

Organotin Oligomeric Drugs Containing the Antiviral Agent Acyclovir

Charles E. Carraher Jr.

Florida Atlantic University, Boca Raton, Florida and Florida Center for Environmental Studies, Palm Beach Gardens, Florida

Robert E. Bleicher

California State University Channel Islands, Camarillo, California

CONTENTS

I.	EARLY HISTORY OF ORGANOTIN COMPOUNDS	76
II.	MECHANISMS AND REACTIONS	76
III.	GENERAL STRUCTURES	77
IV.	ACYCLOVIR	80
V.	BIOACTIVITY OF RELATED COMPOUNDS	81
VI.	EXPERIMENTAL WORK	82
VII.	RESULTS AND DISCUSSION	83
VIII.	REFERENCES	86

Macromolecules Containing Metal and Metal-Like Elements,
Volume 3: Biomedical Applications, edited by Alaa S. Abd-El-Aziz,
Charles E. Carraher Jr., Charles U. Pittman Jr., John E. Sheats, and Martel Zeldin
ISBN: 0-471-66737-4   Copyright © 2004 John Wiley & Sons, Inc.

I. EARLY HISTORY OF ORGANOTIN COMPOUNDS

The initial organotin was reported in 1849 by Franklin.[1] Numerous books have been written about organotin compounds, for instance, Refs. 2–10. Here we will focus on a review of organotin chemistry as related to the compounds described in this study.

II. MECHANISMS AND REACTIONS

While no good mechanistic and kinetic reactions have been reported related to organotin polymer formation from reaction with Lewis bases, there are various related studies with smaller molecules.

The most characteristic feature of organotin halides is their ready ability to undergo nucleophilic substitution as shown below where $Y = RO^-$, R_2N^-, RCO_2^-, and so on and X is a halide.[5,11]

$$Y^- \quad + \quad \begin{matrix} R \\ \backslash \\ R\text{-}Sn\text{-}X \\ / \\ R \end{matrix} \longrightarrow \begin{matrix} R \\ \backslash \\ R\text{-}Sn\text{-}Y \\ / \\ R \end{matrix} \quad + \quad X^- \qquad (1)$$

In fact, the beginning of our work was based on the observation that hydrolysis of group IVA organometallic halides underwent hydrolysis similar to regular organic acid halides. The hydrolysis cycle for mono and dihalo organotin halides is described below.[5]

$$R_3SnX + H_2O \longrightarrow R_3SnOH \longrightarrow R_3Sn\text{-}O\text{-}SnR_3 \qquad (2)$$

$$R_2SnX_2 + \quad H_2O \quad \dashrightarrow \quad \begin{matrix} X \\ / \\ R_2Sn \\ \backslash \\ OH \end{matrix} \quad \longrightarrow \quad \begin{matrix} R & R \\ \backslash & / \\ X\text{-}Sn\text{-}O\text{-}Sn\text{-}X \\ / & \backslash \\ R & R \end{matrix} \quad \longrightarrow$$

$$HO\text{-}Sn(R)_2\text{-}O\text{-}Sn(R)_2\text{-}X \quad \longrightarrow \quad HO\text{-}Sn(R)_2\text{-}O\text{-}Sn(R)_2\text{-}OH \quad \longrightarrow$$

$$\begin{matrix} R \\ | \\ \text{-}(Sn\text{-}O\text{-})_n\text{-} \\ | \\ R \end{matrix} \qquad (3)$$

Hydrolysis or organotin halides does not occur rapidly. In fact, dibutyltin dichloride can be layered over water and left for a day without noticeable hydrolysis occurring. Yet, hydrolysis is rapid when the dibutyltin dichloride is wetted by addition of DMSO, acetone, and similar compounds and then exposed to water. Thus, resistance to hydrolysis is dependent on the hydrophobic behavior of the dibutyltin dichloride rather than on an inherent lack of reactivity. As noted above, when hydrolysis occurs, tetraorganodihalogenodistannoxanes, $XR_2Sn-O-SnR_2X$, are initially formed and then the halide hydroxides, $XR_2Sn-O-SnR_2(OH)$, and finally the so-called polymeric, actually oligomeric, oxides, R_2SnO, are formed.[5]

Once formed, the Sn–O and Sn–N bonds appear to be stable as long as they remain solids. Yet reaction of trimethyltin diethylamine with chloroform and bromoform is reported to occur at room temperature to give the trihalogenomethyltin compounds in good yield.[12,13]

$$Me_3SnNMe_2 \ + HCBr_3 \ +Et_2NH \ \longrightarrow \ Br_3SnNMe_2 \qquad (4)$$

Such compounds also react with acetone to give mesityl oxide and with nitro compounds as 2-nitropropane.[14] Thus, caution must be exercised when attempting to solubilize polymeric products containing polar units in the backbone. It is probable that with many liquids such as DMSO, DMF, DMA, and other dipolar aprotic liquids that complexes are formed, particularly with tetrahedral structures whereby the liquid molecules sit on either side of the tin atom giving a newly formed octahedral structure. It is not known whether this occurs with organotin polymers, but partial solvation of this variety probably occurs.

It is important to note that in the presence of UV light that organotin compounds can degrade. Thus, in the presence of UV light, trimethyltin chloride in carbon tetrachloride degrades into inorganic tin via dimethyltin dichloride and methyltin trichloride.[15]

The reactant compounds, namely, dibutyltin dichloride and dioctyltin dichloride, are commercially produced by the Kocheshkov disproportionation reaction whereby Bu_4Sn (or $Oc_4 Sn$) is heated with $SnCl_4$ at about 240°C for several hours.[16]

III. GENERAL STRUCTURES

In general, tin (IV) compounds are either tetrahedral, trigonal bipyramidal, or octahedral depending on the number and nature of substitutes.[5] For smaller molecules compounds of some combinations of the four substitutes give products that are tetrahedral when the number is five, the products are trigonal bipyramidal, and when it is six, they are octahedral. If the octahedral compound adds another potential binding group this may be violated to include 7- and 8-coordinated tin products.[5]

Salts of acids can chelate through the carbonyl as shown below forming three possible structures.[17] The top two structures, **1** and **2**, are similar to that of a trigonal bipyramid with facial alkyl groups and the two carbon–oxygen groups equivalent. These structures are referred to as bridged structures.

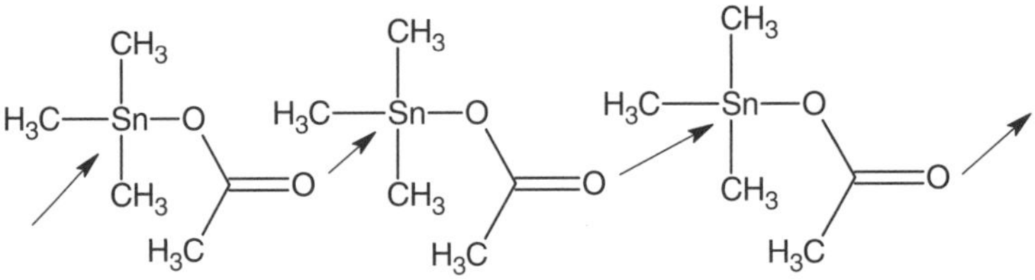

1 (polymeric bridged)

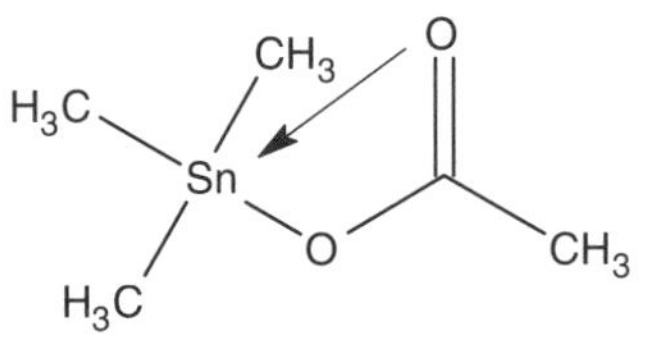

2 (monomeric bridged)

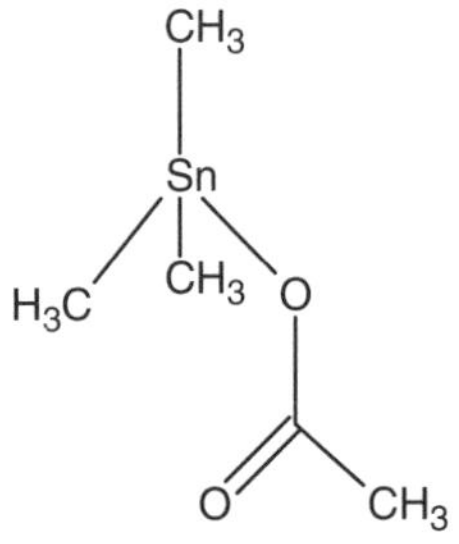

3 (nonbridged structure)

The third structure, **3**, approximates a tetrahedral structure and is referred to as the *nonbridged structure*.

In the solid state, the structures of most compounds of most trialkyltin esters approximate the polymeric bridged structure where the oxygen atoms occupy the apical positions in a somewhat distorted trigonal bipyramid. For example, in trimethyltin acetate, the Sn–O bond distances[18] are 220 and 230 pm and the O–Sn–O bond angle about 172°. For the corresponding triphenyltin acetate the values are 218.5 and 235 pm and 174°.

By comparison, the diorganotin dicarboxylates are monomeric in the solid state but form an octahedral structure.[5,19,20] The structure is similar to that depicted for the monomeric bridged structure, except the arrangement about the tin assumes a distorted octahedral as shown in **4**. The structure as been described as a tetrahedron built from two Sn–C and two short Sn–O bonds that are then capped by weak bonds from the two carbonyl groups.[19] For dimethyltin dibenzoate, the C–Sn–C bond angle is 147° with two short Sn–O bond distances of 216 and 213 pm and two long Sn–O bonds both 251 pm. In solution the ^{17}O NMR spectrum shows only one signal indicating rapid carboxylate group exchange.[20]

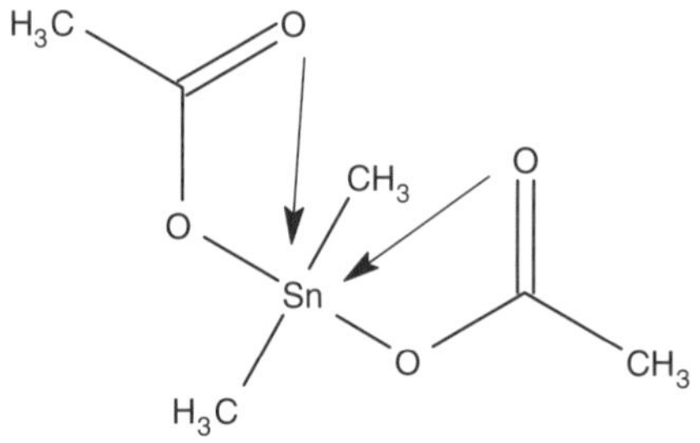

4

For polymers, we have found that the octahedral structure predominates for reactions involving diacid salts.[21-23]

The structures of organotin amines and alkoxides are tetrahedral, as expected. Organotin alkoxides and phenoxides are prepared from reaction with alcohols and alkali metal alkoxides or phenoxides with the organotin halide.[5,24] Most of the published reactions occur in the liquid phase. For instance, the dimethoxide is formed from reaction of dibutyltin dichloride in methanol at 0°C in 98% yield.[25,26]

$$Bu_2SnCl_2 + NaOMe \longrightarrow Bu_2Sn(OMe)_2 + 2NaCl \qquad (5)$$

A similar product can be formed through reaction between dibutyltin oxide and butanol by heating in toluene at 110°C and then heating the resulting dioxide to about 200°C under reduced pressure.[24] However, in tetrahydronaphthalene (boiling point 207°C) phenols give the dialkyltin diphenoxides directly.[27]

$$Bu_2SnO + BuOH \longrightarrow BuO\text{-}Sn(Bu)_2\text{-}O\text{-}Sn(Bu)_2\text{-}OBu \longrightarrow$$
$$Bu_2Sn(OBu)_2 \quad + \quad Bu_2SnO \qquad (6)$$

We have previously synthesized a number of organotin alkoxide polymers employing the reaction of disodiumalkoxides using a room-temperature solution system. The products were high molecular weight materials.[28]

$$R_2SnCl_2 + NaO\text{-}R\text{-}ONa \longrightarrow \overset{R}{\underset{R}{-(\text{-}Sn\text{-}O\text{-}R\text{-}O\text{-})_n\text{-}}} \qquad (7)$$

Similar products were made by employing various modified interfacial systems at room temperature.[29,30] The products were oligomeric.

Like alcohols, amines react with organotin dihalides displacing the halide with the amine-forming amidostannanes. Reaction with a strongly nucleophilic metal amine readily occurs giving the amidostannane product through transmetallation.[31]

$$Me_3SnCl + LiNMe_2 \longrightarrow Me_3SnNMe_2 \tag{8a}$$

The structures of stannylamines are complex and involve an interplay between steric effects, the electronegativities of the nitrogen ligands, and possible π bonding between the nitrogen and tin. The electron diffraction of $(Me_3Sn)_3N$ showed the molecule to be planar about the nitrogen with the Sn−N bond length of 204 pm, but the tin was generally tetrahedral.[32] Except for exceptional cases such as $(ClSnMe_2)_3N$ where, the chloride atoms form bridges between the tin atoms, giving a trigonal bipyramidal structure[33] about the tin, most tin amines are tetrahedral. The ^{15}N NMR or organotin compounds has been reviewed by Wrackmeyer.[34]

Carraher and Winter have synthesized a number of aminostannane polymers employing various interfacial reactions between the diamines and organotin dihalides.[35–38] The products are mainly oligomeric.

$$R_2SnCl_2 + H_2N\text{-}R\text{-}NH_2 \rightarrow \underset{\underset{R}{|}}{-(-N\text{-}R\text{-}N\text{-}Sn\text{-})_n\text{-}} \quad \overset{\overset{H\ \ H\ R}{|\ \ \ |\ \ |}}{} \tag{8b}$$

IV. ACYCLOVIR

Acyclovir,2-amino-1,9-dihydro-9-[(2-hydroxyethoxy)methyl]-6H-purin-6-one, is also known as acycloguanosine and by its tradename Zovirax. It is sold by Glaxo Welcome as a powder, pill, and ointment.

In line with attempts to develop various antibacterial and anticancer agents, Carraher's group are now looking at contributions that can be made with antiviral agents. The approach is to utilize known successful drugs and to incorporate them into polymers where the other "comonomer" can also act to enhance the biological activity. In the present research, the biological activity of organotin moieties is coupled with the biologically active acyclovir. Drugs are created that are active through at least two different routes. This lessens the potential for the microorganism to be able to mutate to nonsusceptible forms and increase the chances that it is effectively inhibited. Further, this approach should discourage the formation of mutated resistant strains.

Acyclovir is a synthetic purine nucleoside that exhibits in vitro and in vivo inhibition of a number of human viruses. In particular, acyclovir is active against herpes simplex types 1 (HSV-1) and 2 (HSV-2) viruses, varicella zoster virus (VZV), Epstein–Barr virus (EBV), and cytomegalovirus (CMV). The inhibitory activity of acyclovir is highly selective. The enzyme thymidine kinase of normal cells does not effectively use acyclovir, but thymidine kinase encoded by one of the viruses noted above converts acyclovir into acyclovir monophosphate. This is converted into the diphosphate by cellular guanylate kinase; finally, it is converted by a number of enzymes into the triphosphate. The triphosphate interferes with herpes simplex virus DNA polymerase and inhibits viral DNA replication. It also inhibits other polymerases. In vitro, the triphosphate can be incorporated into growing DNA chains effectively terminating chain growth. The acyclovir is then less toxic to normal cells because less is taken up, less is converted to the active form, and cellular α-DNA polymerase is less sensitive to the active form. In trials, acyclovir was found to lessen the severity and duration of herpes simplex infections in immunocompromised patients, initial episodes of herpes genitalis, brain biopsy-proven herpes simplex encephalitis, and varicella zoster (shingles) infections in immunocompromised patients. Thus, it is a first line antiviral drug that is also effective at treating a number of other related severe problems.

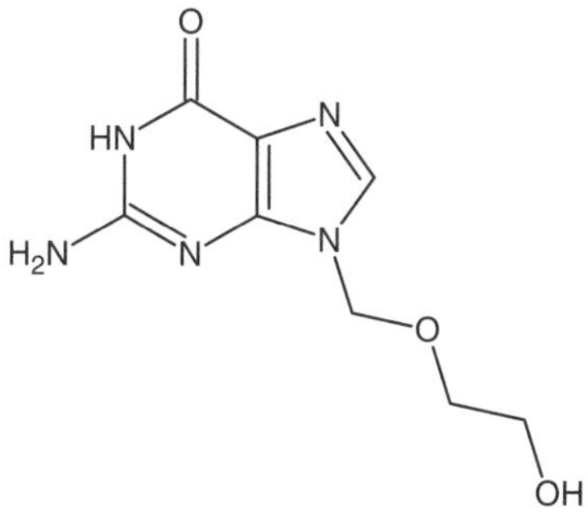

5 (acyclovir)

V. BIOACTIVITY OF RELATED COMPOUNDS

Carraher's group reported the synthesis of a number of organotin-containing polymers for the purpose of synthesizing biologically active materials (e.g., see Refs. 29, 30, 35–47). Many of these were based on the reaction of mono-, di- and trihalo organostannanes with alcohol-containing reactants forming the following linkage:

$$-Sn\text{-}Cl + R\text{-}OH \longrightarrow -Sn\text{-}O\text{-}R \tag{9}$$

These products exhibited a wide variety of biological activities ranging from very specific, particularly when utilizing synthetic alcohol-containing reactants such as poly(vinyl alcohol), to very broad ranging, particularly when employing natural alcohol-containing reactants such as sucrose, dextran, cellulose, and xylan (see, e.g., Refs. 39, 40).

A basic mechanism that is believed to be responsible for this difference in activity involves release of the organotin. In cases where the comonomer is biologically active, release of the comonomer drug occurred through either simple physical hydrolysis or through hydrolysis initiated or is assisted by hydrolyzing enzymes emitted by the particular microorganism. It is uncertain if the polymer, some oligometric fraction of the polymer chain, the monomeric units or some combination of these is responsible for the activity.

As noted before, Carraher has previously reported the synthesis of organotin polyethers[28–30] and organotin polyamines.[35–38] The synthesis of similar materials is reported here except that they are based on the amino alcohol–containing reactant acyclovir. The product will have the following general repeat unit (**6**):

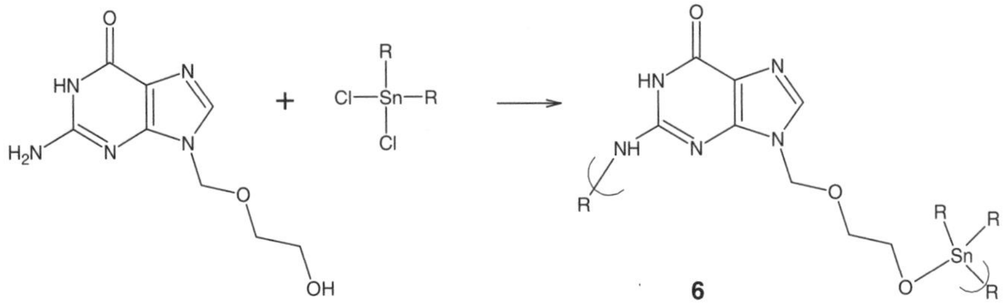

The biological activities of organotin compounds is well known (e.g., see Refs. 10, 29, 30, 35–47). Organotin compounds have long been used in the coatings industry as antifouling, antibacterial, and antifungal agents. They have also been employed as agricultural and horticultural agents against fungal diseases such as early blight, down mildew, anthracnose, and leaf and pod spot on a variety of crops. Organotins have also been developed as pharmaceuticals , anthelmintics, and disinfectants and, more recently, as antitumor drugs. A number of diorganotin compounds have been synthesized as analogues to *cis*-DDP (cis-dichlorodiamineplatinum II), the most widely used anticancer drug. Some of these compounds showed similar, but lower activity against P388 lymphocytic leukemia, but, unlike *cis*-DDP, they exhibited no noticeable nephrotoxicity. It is believed that *cis*-DDP is active because it chelates between amine groups on the DNA, thus modifying the DNA to the extent that it is unable to replicate. Because the bond angles present on a wide variety of both active and inactive organotin compounds are similar, it is believed that the activity of the organotin compounds is due to some other mechanism. Recently, Carraher found that drugs could be developed with good anticancer activity by coupling them with known antibacterial drugs including ampicillin, *p*-aminobenzoic acid, and cephalexin.[21–23,45,46]

VI. EXPERIMENTAL WORK

Chemicals were used as received. Acyclovir [CA (*chem. Abstracts*) 59277-89-3], and dimethyltin dichloride (CA 753-73-1) were purchased from the Aldrich Chemical Company, Milwaukee, WS. Diethyltin dichloride (CA 866-55-7) and dibutyltin dichloride (CA 683-18-1) were obtained from the Peninsular Chemical Research, Gainesville, FL.

The polymers were synthesized employing the classical interfacial technique. Briefly, solutions containing the Lewis base (here 3.00 mmol of acyclovir with 6.00 millimole of sodium hydroxide contained in water) and Lewis acid (here 3.00 mmol of organotin dihalide in carbon tetrachloride) are added with rapid stirring (about 18,000 rpm no load). Reaction is fast (<30 s), and a white solid precipitates from the reaction mixture. The product is washed and allowed to dry under room conditions.

Infrared spectral studies were done utilizing a Mattson Instruments Galaxy 4020 FTIR employing potassium bromide pellets. All spectra were recorded at an instrument resolution of 4 cm^{-1} using 32 scans. Mass spectra were obtained utilizing a HP Mdl. G2025A MALDI-TOF (matrix-assisted laser ionization time-of-flight) mass spectrophotometer.

VII. RESULTS AND DISCUSSION

The reaction is general, giving moderate to good yields (Table 1).

Table 1 Yields for Reaction between Acyclovir and Organotin Dihalides[a]

Organotin Dihalide	Yield (g)	Yield (%)
Dibutyltin dichloride	1.26	92
Dimethyltin dichloride	1.00	90
Diethyltin dichloride	0.28	19
Dioctyltin dichloride	1.18	69
Diphenyltin dichloride	1.19	80
Dicyclohexyltin dibromide	1.85	120[b]

[a]*Reaction conditions*: Diorganotin dihalide (3.0 mmol) dissolved in 30 ml of heptane is added, at room temperature (~25°C) to a rapidly stirred (18,500 rpm no load) solution of acyclovir (3.0 mmol) and sodium hydroxide (6.0 mmol) dissolved in 30 mL of water and run for about 10 s.

[b]Probably has excess bromide end groups.

Infrared spectral results are consistent with the presence of moieties derived from the organotin and acyclovir. A brief discussion of some of the assignments follows. All infrared assignments are given in cm^{-1} and literature assignments are given elsewhere.[29,30,35–47] The product from dimethyltin and dichloride and acyclovir will be described. Bands about 3100 are assigned as C–H aromatic from the acyclovir. Bands at about 2715 are assigned to the C–H aliphatic stretch from the aliphatic ether arm on the acyclovir (the "C" unit as described below). Bands at about 2540 are assigned to C–H stretching of methyl groups on the dimethyltin moiety. The band at about 1700 is assigned as derived from the purine ring. Purines exhibit several characteristic aromatic ring vibrations found at about 1630 and 1480. A band at 795 is prominent in both the spectra of the product and dimethyltin dichloride but not in acyclovir. A band at about 1460 is assigned to the Me–Sn stretch. The Sn–Cl is found in the range of 315–400 below the range of the instrument used to record the spectra. In acyclovir there is a much broader band(s) from 3500 to 3170 that includes both the N–H and O–H stretching vibrations. For the product, much of the 3500–3170 band(s) is missing consistent with the formation of the Sn–O and absence of the –OH grouping. There is a band about 3445 that is assigned to the N–H stretch. Bands in the 3700 region are probably due to the presence of some Sn–OH end groups.

The Sn–N band is assigned as occurring at about 1100. For the product with dimethyltin dichloride, this band appears at about 1120. A similar band is found for some of the other products. Where it is not found, a large band at about 1100 appears to hide it. The Sn–O band is assigned to be within the range of 400–700. A new band at about 420 is tentatively assigned to be due to the presence of the Sn–O linkage. Thus, the linkage of the dialkyl or diaryltin moiety with the acyclovir through the nitrogen and oxygen is indicated by the infrared spectra.

Many of the condensation metal-containing polymers are poorly soluble for a number of reasons. Thus, it is difficult to obtain accurate molecular weights for these materials. Further, many have good thermal stabilities, often with the metal remaining to higher than 1000°C, making accurate elemental analysis difficult. Even so, mass spectral analysis is often able to allow the identification of one or more repeat units, consistent with the formation of at least oligomers. We have begun investigating the use of MALDI-TOF mass spectroscopy as a tool to allow a better structural identification of at least some fraction of the product. True MALDI requires that the polymer is soluble in some low boiling solvent that allows more intimate mixing and dispersion of the sample compound with the laser sensitive matrix liquid. The present products are only slightly soluble in DMSO, which is not considered a volatile liquid for MALDI. (The search is continuing for different and better solvents.) In the present study we ground the sample compound with the matrix liquid, giving a finely dispersed mixture. The mixture was introduced to the instrument and MS obtained in the usual fashion. The following results are preliminary.

Seemingly, the best spectra and highest ions were observed for the dioctyltin product. Results appear in Table 2 for significant ions with molecular masses greater than 400 daltons. (For each of the mass spectral data tables the following conventions will be used. First, A, 7, signifies the dioctyltin moiety; B represents the acyclovir moiety; and C represents the diethyl unit, namely, –CH₂–O–CH₂–CH₂–O–.) End groups were assigned for many of the products. End-group assignment is often difficult and even identification hard. MALDI appears to allow ready identification for some of the ions. With respect to products possessing tin end groups, two end groups are probable—the hydroxyl resulting from the hydrolysis of the tin chloride, and the chloride (Sn–OH and Sn–Cl). Both end groups are found. Little fragmentation occurs for the higher-mass ions, with the most readily observable leaving groups which are the diether linkage designated as "C" and octyl groups. In this case, oligomeric chains are identifiable, employing MALDI along with the assignment of end groups associated with the terminal tin moiety.

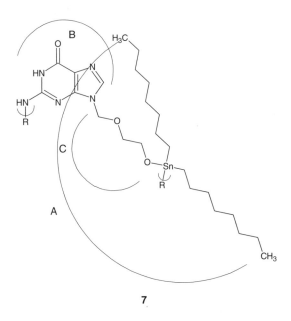

7

Table 2 Mass Spectral Results for the Product of Dioctyltin Dichloride and Acyclovir for $m/e > 400$

m/e	(Possible) Assignment	m/e	(Possible) Assignment
3456	6 units with Cl end group	1099	Cl-ABAB minus C
3340	6 units minus C	1012	ABAB-Octyl(?)
2088	3 units + A with Cl end group	835	Cl-ABA minus octyl
1745	3 units + Cl end group	455	AB minus octyl
1154	2 units + OH end group		

Table 3 contains similar results for the product from diphenyltin dichloride and acyclovir. Again, ion fragment results are consistent with the structural assignment.

Table 3 Mass Spectral Results for the Product of Diphenyltin Dichloride and Acyclovir for $m/e > 400$

m/e	(Possible) Assignment	m/e	(Possible) Assignment
1490	3 units	571	BAB minus 2C
1425	3 units minus C	501	AB
1000	Cl-ABA-Cl	442	HO-AB minus C
790	ABA		

In summary, oligomeric organotin products have been synthesized from the reaction between organotin dihalides and acyclovir with bonding occurring through Sn–O– and Sn–N–.

VII. REFERENCES

1. E. Franklin, *J. Chem. Soc.* **2**, 263 (1849).

2. W. P. Neumann, *The Organic Chemistry of Tin*, Wiley, New York, 1970.

3. A. K. Sawyer, *Organotin Compounds*, Marcel Dekker, New York, 1971.

4. R. C. Poller, *The Chemistry of Organotin Compounds*, Logos Press, London, 1970.

5. A. G. Davies, *Organotin Chemistry*, VCH, New York, 1997.

6. J. Zuckerman, *Organotin Compounds: New Chemistry and Applicaitons*, ACS, Washington, DC, 1976.

7. T. Sato, *Main-Group Metal Organometallics in Organic Synthesis: Tin*, Pergamon, Oxford, 1995.

8. P. Harrison, in *Dictionary of Organometallic Compounds*, ed., J. Macintyre, Chapman & Hall, London, 1995.

9. S. Patai, *The Chemistry of Organic Germanium, Tin, and Lead Compounds*, Wiley, New York, 1995.

10. I. Omae, *Organotin Chemistry*, Elsevier, Amsterdam, 1989.

11. J. Spencer, T. Ganuis, A. Zafar, H. Eppey, J. C. Otter, S. Coley, C. Yoder, *J. Organomet. Chem.* **389**, 295 (1990).

12. A. G. Davies, T. Mitchell, *J. Chem. Soc.* (*C*) **1896** (1969).

13. T. Chivers, B. David, *J. Organomet. Chem.* **13**, 177 (1968).

14. K. Jones, M. Lappert, *J. Organomet. Chem.* **3**, 295 (1965).

15. S. Blunden, *J. Organomet. Chem.* **248**, 149 (1983).

16. J. G. A. Luijten, G. J. M. van der Kerk, *Investigations in the Field of Organotin Chemistry*, Tin Research Institute, Greenford, 1955.

17. E. Tiekink, *App. Organomet. Chem.* **5**, 1 (1991).

18. K. Molloy, T. Purcell, K. Quill, I. Nowell, *J. Organomet. Chem.* **267**, 237 (1984).

19. E. Tiekink, *J. Organomet. Chem.* **408**, 323 (1991).

20. A. Lycka, J. Holecek, *J. Organomet. Chem.* **294**, 179 (1985).

21. C. Carraher, H. Leuing, *Polym. Mater. Sci. Eng.* **82**, 81 (2000).

22. C. Carraher, F. Li, C. Butler, *J. Polym. Mater.* **17**, 377 (2000).

23. C. Carraher, F. Li, *Polym. Mater. Sci. Eng.* **83**, 405 (2000).

24. A. Bloodworth, A. G. Davies, *Organotin Compounds*, in A. Sawyer, ed., Marcel Dekker, New York, 1971.

25. D. Alleston, A. G. Davies, *J. Chem. Soc.* 2050 (1962).

26. J. Kennedy, W. McFarlane, P. Smith, R. White, *J. Chem. Soc., Perkin Trans.* **2**, 1785 (1971).

27. A. G. Davies, D. Kleinschmidt, P. Palan, S. Vasishtha, *J. Chem. Soc.* (*C*) 3972 (1971).

28. C. Carraher, G. Scherubel, *Makromol. Chem.* **160**, 259 (1972).

29. C. Carraher, G. Scherubel, *Makromol. Chem.* **152**, 61 (1972).

30. C. Carraher, G. Scherubel, *J. Polym. Sci. A-1*, **9**, 983 (1971).

31. K. Jones, M. Lappert, *J. Chem. Soc.* 1944 (1965).

32. L. Khaikin, A. Belyakov, G. Kooptev, A. Golubinskii, L. Vilkov, *J. Mol. Struct.* **66**, 191 (1980).

33. C. Kober, J. Kroner, W. Storch, *Angew. Chem., Int. Ed. Engl.* **32**, 1608 (1993).

34. B. Wrackmeyer, E. Kupce, *Topics Phys. Org. Chem.* 4 (1992).

35. C. Carraher, D. Winter, *Makromol. Chem.* **152**, 55 (1972).

36. C. Carraher, D. Winter, *Makromol. Chem.* **141**, 259 (1971).

37. C. Carraher, D. Winter, *J. Macromol. Sci.-Chem.* **A7**(6), 1349 (1973).

38. C. Carraher, D. Winter, *Makromol. Chem.* **141**, 237 (1971).

39. C. Carraher, D. Giron, J. Schroeder, C. McNeely, U.S. Patent 4,312,981 (1982).

40. C. Carraher, C. Butler, U.S. Patent 5,840,760 (1998).

41. C. Carraher, Butler, Y. Naoshima, D. Sterling, V. Saurino, *Biotechnology and Bioactive Polymers*, Plenum, New York 1994.

42. C. Carraher, J. Piersma, *Angew. Makromol. Chem.* **28**, 153 (1973).

43. C. Carraher, L. Reckleben, C. Butler, *PMSR* **63**, 704 (1990).

44. C. Carraher, M. Fedderson, *Angew. Makromol. Chem.* **54**, 119 (1976).

45. C. Carraher, F. Li, C. Butler *J. Polym. Mater.* **17**, 377 (2000).

46. C. Carraher, F. Li., *Polym. Mater. Sci. Eng.* **83**, 405 (2000).

47. C. Carraher, *Angew. Makromol. Chem.* **31**, 115 (1973).

CHAPTER 6

Polymeric Ferrocene Conjugates as Antiproliferative Agents

Eberhard W. Neuse

School of Chemistry, University of the Witwatersrand, Johannesburg, South Africa

CONTENTS

I. INTRODUCTION   90

II. THE FERROCENE–FERRICENIUM SYSTEM IN THE
BIOLOGICAL ENVIRONMENT   92

III. POLYMER–DRUG CONJUGATION AS A PHARMACEUTICAL TOOL
FOR DRUG DELIVERY   98

IV. POLYMER–FERROCENE CONJUGATES:
SYNTHESIS AND STRUCTURE   100
   A. The Carrier Component: Structural Considerations   101
   B. Conjugates of Amide-Linked Ferrocene   102
   C. Conjugates of Ester-Linked Ferrocene   109

V. BIOACTIVITY SCREENING   110

VI. SUMMARY AND CONCLUSIONS   113

VII. ACKNOWLEDGMENTS   115

VIII. REFERENCES   115

Macromolecules Containing Metal and Metal-Like Elements,
Volume 3: Biomedical Applications, edited by Alaa S. Abd-El-Aziz,
Charles E. Carraher Jr., Charles U. Pittman Jr., John E. Sheats, and Martel Zeldin
ISBN: 0-471-66737-4   Copyright © 2004 John Wiley & Sons, Inc.

I. INTRODUCTION

Organometallic and metal coordination compounds represent a broad class of metal-containing molecules in which the metal atom is either directly connected to a C atom of an organic moiety or coordinatively bonded to heteroatoms of organic or inorganic donor ligands. Metals participating in these types of bonding may be maingroup elements or members of the transition metal series. Being in atomic dispersion, such metal centers offer challenging opportunities in both fundamental and applied research because of the intriguing physical and chemical features[1-3] played out by such centers in the technological and biological environments.

Of particular interest are the metallocenes, a class of organometallic compounds in which a metal atom is π-bonded, generally in p_π–d_π fashion, to one or two aromatic ring structures of the cyclobutadiene, cyclopentadienyl, or benzene types as exemplified in Figure 1.

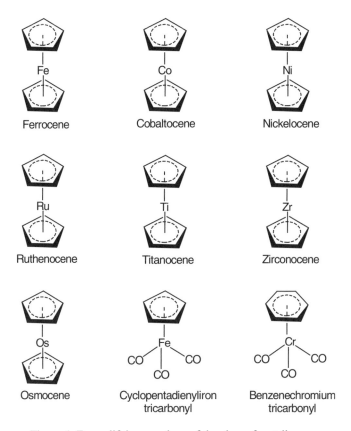

Figure 1 Exemplifying members of the class of metallocenes.

The organoiron compound, ferrocene, di-(η^5-cyclopentadienyl)iron(II) (**1**), stands out as a unique example of this class. Ever since its discovery[4] in 1951, it has followed an illustrious career path stretching from basic chemistry and physics to numerous branches of technology and, largely since the early 1980s, of biology and medicine. The principal reason for the diversified activity spectrum of ferrocene can be found in its unusual oxidation–reduction behavior. Removal of an electron from an essentially nonbonding HOMO (highest occupied molecular orbital) readily converts the neutral compound to its one-electron oxidation product, the ferricenium cation **1**⁺, an ion radical of unusual stability.[1,5] The process is reversible, one-electron reduction regenerating the parent compound (Scheme 1), and the inherent electron transfer has considerable ramifications in such diverse technical fields as flame retardance, photoresist and electrode technologies, catalysis, and, demonstrated in the most spectacular form, in solid rocket propulsion, where some of its monomeric derivatives, as well as selected polymeric compounds,[6] served as "ballistic modifiers" in the heydays of cold-war competition in space technology.

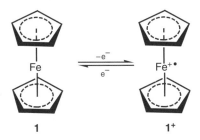

Scheme 1

Moving on to more recent years, one should not be surprised to find ferrocene research to make ever growing inroads into the realm of the biosciences. (For an early review, see Dombrowski et al.[7]). Free-radical reactions play an important role in numerous biological and biomedical processes,[8] and the ferrocene–ferricenium system, once introduced into the biological environment, should find willing partners in all reaction steps involving electron transfer. The material surveyed in the subsequent section, reflecting on that topic, will provide the background information underlying the motivation for the ferrocene polymer investigations discussed in the main Sections IV and V.

II. THE FERROCENE–FERRICENIUM SYSTEM IN THE BIOLOGICAL ENVIRONMENT

Oxidation–reduction and free-radical reactions are vital links in the complex chemical network of biological processes. The redox behavior and electron transfer properties of ferrocene, alluded to in the introductory section, should, hence, provide an intriguing proving ground for research in the biochemical and biomedical domains. The literature indeed reveals a large number of studies concerned with ferrocene and its oxidized counterpart in the biological environment. It has been found, for example, that the enzyme-mediated reaction of ferrocene with hydrogen peroxide generates the ferricenium cation;[9] the oxidant may biologically arise from glucose by the action of glucose peroxidase. The cation **1**[+], in turn, may undergo a variety of transformations. Thus, it has the potential for charge transfer complex formation by reaction with π-electron donor groups, such as tryptophan, in proteins. [10] In addition, and more importantly, it may scavenge other, potentially aggressive, free radicals through recombination,[1] leading to substituted ferrocenes after proton elimination (Scheme 2).

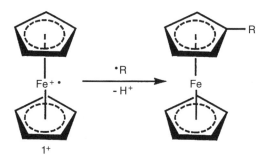

Scheme 2

The cation is, of course, susceptible to reduction reactions. It will be reduced, for instance, by metalloproteins, such as cytochrome C or plastocyanin.[11,12] Particularly significant is its reduction by NADH.[13] As a coenzyme for dehydrogenases, NADH plays a vital role in the control of biological redox systems. Oxidized by **1**[+], it becomes transformed into the radical cation NADH[+]·, which in turn, in the presence of base, converts to the neutral free radical NAD· and, further, to the NAD[+] cation, with **1**[+] involved as an electron acceptor in the last-named reaction step and **1** regenerated. This sequence of events, illustrated in Scheme 3, demonstrates the capability of oxidized ferrocenes to interfere rather drastically in enzymatically controlled electron transfer reactions.

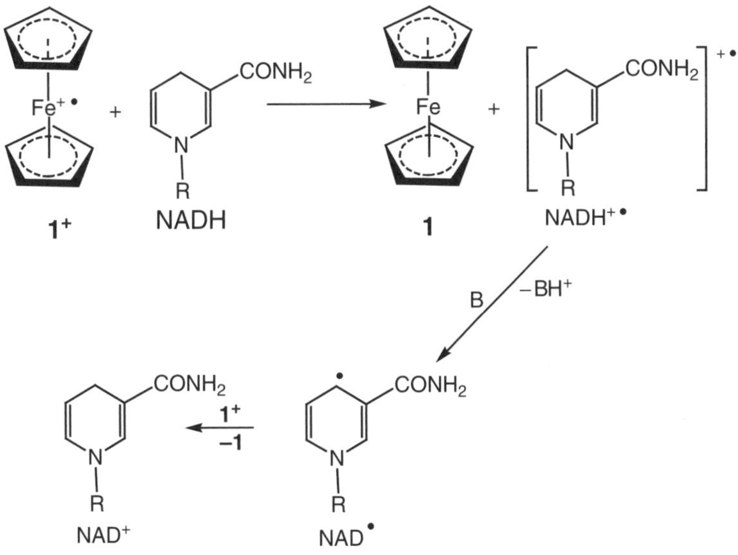

R = adenosine diphosphoribosyl

Scheme 3

A further ferricenium reduction process, investigated by Espenson's group,[14] is mediated by the biologically important, flash-photolytically generated superoxide anion radical, which in the process is transformed into dioxygen (Scheme 4).

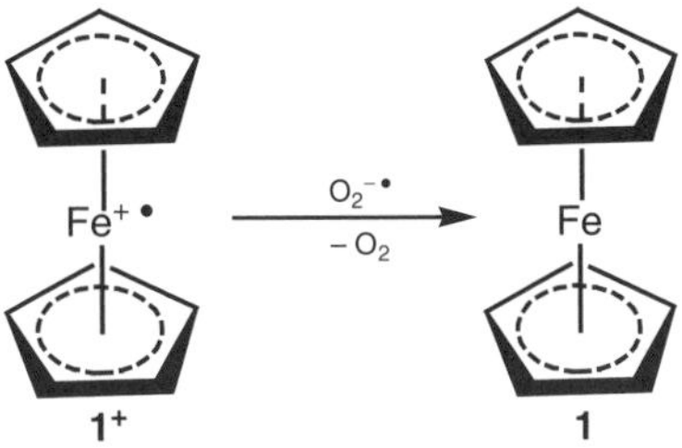

Scheme 4

A related superoxide scavenging reaction, this one studied by Logan,[15] involves a zwitterion-type ferricenylalkylcarboxylate, $Fc^+-(CH_2)_3-COO^-$, which is reduced to the corresponding ferrocenylalkylcarboxylate, $Fc-(CH_2)_3-COO^-$, again with concomitant formation of dioxygen (Scheme 5).

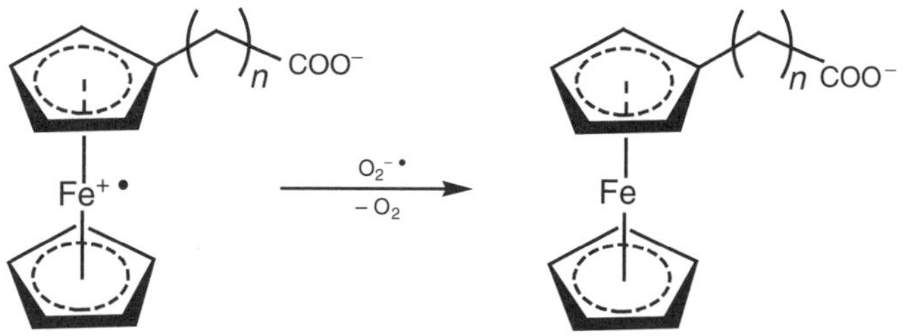

Scheme 5

Of particular interest is Logan's observation[16,17] that ferrocenylalkylcarboxylates of the very same type, when photooxidized in the presence of N_2O, convert to the ferricenyl counterparts with concomitant generation of the hydroxyl radical, and the latter immediately attacks and oxidizes another molecule of the ferrocenyl derivative while being transformed itself into hydroxyl anion (Scheme 6).

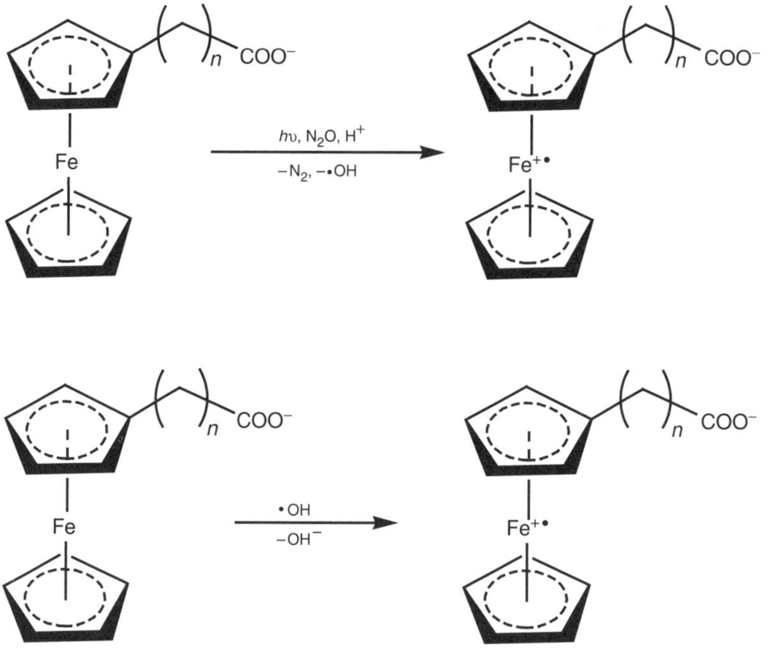

Scheme 6

Here, then, the ferrocenyl compounds act as hydroxyl radical scavengers. On the other hand, as depicted in the simplified Scheme 7, ferricenium salt, exemplified by ferricenium hexafluorophosphate, undergoes reduction to ferrocene in the presence of glutathione (GSH)[18] with simultaneous formation of the very same hydroxyl radical that was scavenged in the preceding reaction sequence. A more recent study confirms the feasibility of ·OH generation from ferricenium salt under physiological conditions.[19] It is in fact suggested in that work[19] that the hydroxyl free radical so generated might be the species instrumental in the cell growth-inhibiting activity of ferricenium salts, leading to oxidative damage of the cellular DNA.

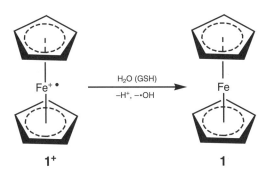

Scheme 7

The findings described above reflect the enormous variability of redox processes—some of these, indeed, seemingly at odds with each other—that will arise in a complex biological system with the ferrocene–ferricenium couple as its core. It was this wide-ranging behavior pattern of the ferrocene and ferricenium compounds which, in the early 1980s, prompted the initiation of a research program in the author's laboratory with the aim of probing the potential usefulness of ferrocene in its oxidized (i.e., free-radical) form as an antiproliferative agent in the fight against cancer.

Free radicals are major role players in the complex biochemical sequences of reactions involved in carcinogenicity and in the various stages of growth and control of malignancies.[20–22] In normal tissue a critical role is played by superoxide dismutase (SOD). This enzyme dismutates superoxide anion into H_2O_2 and O_2, and the former is eliminated by the action of catalase and glutathione peroxidase, thus preventing its further transformation by a Fenton-type mechanism into the deleterious hydroxyl radical. Most transformed (i.e., cancerous) cells, in contrast, are characterized by considerably reduced SOD activity, and the superoxide levels are correspondingly high. Superoxide and secondary radical species may bind to the target cell's nuclear DNA, thereby causing replication errors during the mitotic stage, and this has been implicated as one of the causative factors in malignancy.[21] Set against that, the metabolism of certain antitumor agents involves free-radical chemistry,[23,24] and their biological activity may well, or in part at least, be associated with some form of intervention via free-radical species. It is known also that certain antioxidants and compounds acting as radical scavengers, including SOD-mimetic metal compounds, can inhibit carcinogenic[22,25] and even metastatic[26] processes. Ferricenium salts might indeed fit well into this broad category of compounds.

With this conception in mind, admittedly rather nebulous at the time, we set out to synthesize a selected number of ferricenium salts[5] and, in collaboration with overseas biomedical laboratories, had these screened for antineoplastic activity against murine tumors (implanted Ehrlich ascites[27,28] and Yoshida ascites[29]) and against several series of human tumor clonogenic cultures.[30] The findings of these early studies revealed lack of activity for the ferrocene parent and a ferricenium salt containing the bulky and hydrophobic heptamolybdate anion, both compounds characterized by very poor solubility in aqueous media. In contrast, most encouragingly, all those ferricenium salts tested that were distinguished by water solubility possessed good to excellent activity. In the Ehrlich ascites tests,[27] for example, both ferricenium picrate and the solvated trichloroacetate, while of very low toxicity (LD_{50} ~ 340–400 mg/kg), provided a cure rate of 100%. Ferricenium salt activity was confirmed in the clonogenic assay,[30] which also, significantly, revealed some activity (after an incubation period) even for the unoxidized but moderately water-soluble ferrocenylacetic acid. Following a comprehensive synthesis program in the author's laboratory to provide a larger number of compositionally well defined ferricenium salts in the state of purity required for biological assays,[5] continued screening work in one of the partner institutions[31] involved testing against several solid murine tumors. Moderate to excellent tumor growth inhibition was observed with all salts tested; the trichloroacetate and picrate salts, again, proved to be the top performers, causing tumor mass reduction (relative to untreated controls) by 60–70% against Lewis lung and, more illuminating still, by 50–70% against the Colon 38 carcinoma, a murine tumor with very restricted response to common cytostatic drugs. Particularly rewarding were the results obtained in tests against xenografted human tumors, notably the L261 lung adenocarcinoma and the M3 breast carcinoma, where the leading picrate and trichloroacetate salts, in the protocols used, caused tumor reduction by approximately 50–70%. The same approximate range was determined for implanted R85, a human colorectal carcinoma known to show considerable resistance to chemotherapy by most common drugs.

In subsequent years, confirmatory evidence was provided from other laboratories for the inactivity of ferrocene and several aminoalkylated ferrocenes against a number of murine tumors[32] and, conversely, for the activity of various ferricenium compounds against murine cell lines and the CH1 human ovarian line.[33] In the CH1 tests, for example, the unsubstituted ferricenium tetrachloroferrate gave an IC_{50} value as low as 10 μM, and, quite noteworthy, the same IC_{50} value was determined for a related, yet nonoxidized metallocene of the bis(hexamethylbenzene)iron(II) type. In a Russian laborary[34,35] high activity was also established for ferricenium and 1,1′-diethylferricenium triiodides, the last-named compound giving a therapeutic index as high as 7.1 (to be compared with 2.3 for the time-proven anticancer drug cisplatin, included in that screen). The very same salt, interestingly, showed no increase in lifespan in tests performed by the Russian group[34] against implanted MCH-11, a methylcholanthren-induced sarcoma, well underlying the time-proven fact that the responses elicited by a given drug from dissimilar neoplasias tend to vary over a wide range.

Up to this point, the findings reported in various laboratories suggest that activity in general is shown only by the cationic ferricenium species, yet not by the unoxidized ferrocene parent. Where activity was observed with ferrocene compounds proper, such as with the ferrocenylacetic acid in the cited clonogenic tests,[30] metabolic oxidation to the corresponding ferricenium counterparts could be expected. This could plausibly also be the case with a glucose derivative containing ferrocene units attached *via* dicarboxyl links, which was synthesized in Keppler's laboratory[36] and found to be active against the CH1 tumor ($IC_{50} = 0.34$ mM). Here, we may speculate that metabolic oxidation of the glucose part (conceivably mediated by glucose peroxidase) might be accompanied by electron transfer from the ferrocene system, the latter acting as a mediator being transformed in the process to the bioactive ferricenium salt. Similarly, the series of ferrocenylalkylated benzotriazole derivatives studied in the aforementioned Russian laboratory[35] and found to possess antitumor activity, could well have been metabolically preoxidized in vivo to the ferricenium stage (or, alternatively, through protonation of a triazole nitrogen atom, be converted to a cationic compound with the ferrocene unit left unchanged), as the authors themselves speculated, before the biological activity manifested itself. Conflicting results have been reported by the Russian group[34] for a number of bis-ferrocenylalkylated benzotriazoles occurring in the cationic state with the positive charge on the heterocycle. No potential metabolic transformation to corresponding ferricenium compounds was indicated in that work. The authors had observed, rather disturbingly, that under certain conditions these compounds, initially antitumor-active, subsequently began promoting tumor generation. From this fact they concluded that the compounds exerted their activity while still in the original molecular state, perhaps reacting with active centers of DNA nucleotides, and that only on gradual catabolic transformation involving oxidation of the ferrocene units and ultimate release of ionic iron was the opposite effect of tumor development acceleration prompted. Obviously more work is needed to shed light on this controversial topic.

Irrespective of the actual mechanism underlying the tumoristatic action, the prevailing conclusion to be drawn from the collected experimental data must be that any ferrocene compound qualifying for candidacy as an antitumor drug should be water-soluble and either exist directly in a cationic (i.e., oxidized ferricenium-type) dosis form or else possess the capacity for smooth in vivo transformation into that oxidized state on parenteral administration.

We drew attention earlier[29] to the inherent instability of the ferricenium cation in aqueous solution at the physiological pH (7.3). When administered intravenously or intraperitoneally into the aqueous body fluid system, a ferricenium compound would thus be expected to possess a half-life too short for predominant survival en route to the target tissue, a major portion of the salt converting back to the uncharged ferroene derivative in the process. However, the activity of the (nonoxidized) ferrocenylacetic acid observed in the clonogenic tests,[30] which, significantly, manifests itself only after some time lapse, leads to the conclusion that a ferrocene-type drug may not necessarily have to be administered in the oxidized, namely, ferricenium, state. We have argued[37,38] that the equilibrium concentrations of the ferrocene and ferricenium species in a particular body compartment will not depend on the compound's initial oxidation state but will, rather, be controlled by the specific physiological conditions prevailing in that compartment, such as pH and oxidizing, reducing, or otherwise metabolizing, enzyme action. It should, therefore, be immaterial whether a compound enters the body's fluid system in the reduced (ferrocene) or oxidized (ferricenium) form, provided only that it possesses sufficient solubility in the aqueous phase for smooth dissolution and distribution.

A strategy designed to achieve ferrocene administration in a water-soluble dosis form could logically involve conversion of the compound to a prodrug constructed so as to possess water solubility and sufficient stability in circulation to survive transport to, and into, the cancerous cell. There the original compound would be released for action, rapidly undergoing the particular ferrocene–ferricenium equilibrium distribution dictated by the intracellular (lysosomal) environment. This prodrug concept can indeed be reduced to practice through the expediency of conjugating the ferrocene unit bioreversibly to a suitable carrier that is polymeric, is water-soluble, and possesses the necessary chemical and physical prerequisites prescribed by the rules of biomedicine. The modus operandi of this carrier-drug conjugation approach will be delineated in the following section, preparing the reader conceptually for the chapter's main topic to be presented in Sections IV and V.

III. POLYMER–DRUG CONJUGATION AS A PHARMACEUTICAL TOOL FOR DRUG DELIVERY

Cancerous diseases, together with cardiovascular afflictions, are presently considered to be the two most efficient killer diseases affecting humans in all parts of the globe, although within wide variability limits. Their impact is further enhanced by burgeoning urbanization, particularly in the developing world, which adds to the environmental and dietary conditions conducive to carcinogenesis, and this is compounded by the rapid spread of AIDS-related afflictions, immuno-depressed patients being at special risk to develop early cancerous lesions. While, in selected cases, the treatment of neoplastic afflictions with medicinal agents, alone or in combination with other modalities, has been undisputedly successful in retarding uncontrolled proliferation and even achieving complete regression, much room is left for improvement in present-day chemotherapy as a tool in the fight against cancer.

Currently administered anticancer drugs, even the most time-honored ones, are fraught with pharmacological deficiencies. Thus, many drugs are salt-like or polar and, therefore, encounter difficulties in membrane crossing and cell entry by the common passive diffusion mechanism with resultant low intracellular bioavailability. Other agents may be poorly soluble in aqueous media; hence, they are inefficiently dissipated in the vascular system and so expose themselves to attack and elimination by the reticuloendothelial system. Most agents lack cell specificity and thus show affinity to both normal and cancerous body tissues; this leads to massive dilution and poor accumulation in the affected cells. Short serum half-lives, associated with rapid enzymatic degradation, protein binding, or excretion through the glomerular system, are a common feature of anticancer agents. The therapeutic window is generally narrow, and the spectrum of drug activity seldom extends over a wide range of different cancers. Quite common also are severe, mostly dose-limiting toxic side effects and a tendency to induce drug resistance, both of these features invariably necessitating treatment discontinuation before regression-free cure rates are achieved. A serious restriction of therapeutic effectiveness is the net effect of the described shortcomings.

While the focus of advanced pharmacological research will always be on the development of new drugs, the design of efficacious drug delivery modalities capable of overcoming the unsatisfactory performance pattern of currently used anticancer drugs is emerging as a more immediate and quite pressing task. An obvious delivery system would be based on some form of prodrug·designed for intact extravasation from the blood pool and transportation of the pharmacologically active component to the target tissue.

One of the most promising strategies designed to translate this prodrug concept into a practical treatment modality involves the conjugation, that is, the temporary bioreversible interconnection, of a drug molecule with a polymeric carrier. This technology thus takes advantage of the pharmacokinetic peculiarities of polymeric compounds. As a macromolecular species, a polymer, when administered parenterally in a water-soluble form, can utilize intercellular and cell entry mechanisms more consistently, and often, more efficaciously, than is generally observed with nonpolymeric compounds. In particular, the EPR (enhanced permeation and retention) effect described by Maeda[39] allows macromolecular compounds uniquely to accumulate in tumorous tissue as a result of vascular leakage and lack of lymphatic drainage. The polymer–drug conjugate, as first proposed in principle by Ringsdorf,[40] consists of a linear polymer chain composed of a major proportion of subunits bearing hydrosolubilizing groups, a minor proportion of subunits bearing the bioreversibly bound drug, and in the ideal case also some subunits with special affinity for the (cancerous) target tissue. The water-soluble carrier component in this assembly functions as a vehicle transporting the medicinal agent, sterically protected from enzyme attack, protein binding, and other scavenger action, through various body compartments and membranes to, and into, the lysosomal compartment of the affected, that is, neoplastic cells. Here the agent is enzymatically or hydrolytically released in monomeric form for biological action. This mode of drug transport, schematically depicted in Scheme 8 for a cytotoxic drug conjugate, profoundly lowers free drug concentration in the vascular system and thereby reduces systemic toxicity. The conjugate's cell entry follows a pinocytotic mechanism,[41] thus circumventing potential hindrance by drug polarity or ionicity, and increased translocation efficacy is observed in the special case of adsorptive pinocytosis as typically realized with cationic polymers.[42] Where excessive P-glycoprotein-mediated efflux of the agent from endocytic space may have developed, the pinocytotic pathway of cell entry assumes outstanding importance as a means of neutralizing the much dreaded phenomenon of drug resistance. It is clear from the foregoing arguments that the net benefit of drug conjugation to a suitably constructed carrier polymer will be an appreciable enhancement of therapeutic effectiveness. For a discussion of exemplifying polymer–drug conjugate syntheses and biological effects, involving a great variety of medicinal agents, the reader is referred to two recent reviews[43,44] as well as an instructive earlier one.[45]

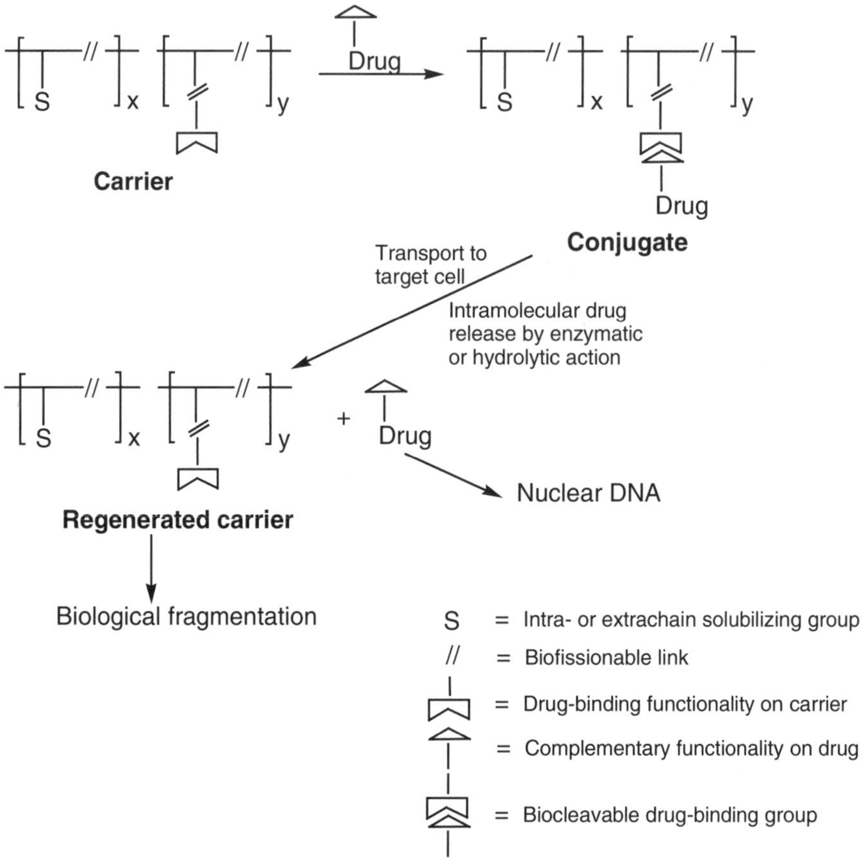
Drug
Carrier
Drug
Conjugate
Transport to
target cell
Intramolecular drug
release by enzymatic
or hydrolytic action
Drug
Nuclear DNA
Regenerated carrier
Biological fragmentation
S = Intra- or extrachain solubilizing group
// = Biofissionable link
= Drug-binding functionality on carrier
= Complementary functionality on drug
= Biocleavable drug-binding group

Scheme 8

IV. POLYMER–FERROCENE CONJUGATES: SYNTHESIS AND STRUCTURE

In Section II we discussed fundamental investigations, reported from various laboratories, of the behavior of ferrocene and its oxidation product, the ferricenium cation, in the biological and biomedical realms, notably in areas concerned with cancer research. We concluded that a promising, water-soluble dosis form for in vivo administration of the ferrocene–ferricenium couple could conceivably be obtained by conversion of the hydrophobic, poorly water-soluble ferrocene complex to a prodrug assembly as represented by a water-soluble conjugate containing the metallocene in a bioreversibly polymer-anchored mode. Section III provided a background to that strategy, outlining the basic rationale for, and the biomedical benefits to be derived from, such polymer anchoring of the primary drug species. In the present section it will be shown how polymer–drug conjugation has been used in practice for the synthesis of carrier-anchored ferrocene.

A. The Carrier Component: Structural Considerations

During the three decades following the discovery of ferrocene in 1951, the literature cited an extraordinarily large number of publications in the area of polymeric ferrocene compounds, as scientists embarked on a search for new properties and applications of the metallocene that would be specifically associated with the polymer-bound, yet not with the monomeric, compound.[46,47] A special type of polymer used in space technology, a linear construct in which ferrocene nuclei are interconnected by simple methylene bridges,[48,49] represents just one example of these early endeavors. Whenever linear and essentially free from rotational restrictions, these polymeric, and quite lipophilic, ferrocenes were generally reported to possess solubility in a restricted number of organic solvents, yet certainly not in water.

In order to meet the vital requirement of water solubility for biomedically useful conjugates, evidently, new strategies profoundly deviating from these earlier ferrocene polymer investigations are needed. The successful synthesis of polymeric ferrocene conjugates dissolving smoothly in aqueous media is based on the important precondition that the carrier component itself be water-soluble and, more importantly still, possess sufficient "carrying power" to maintain solubility on conjugation with the hydrophobic metallocene. The exposé in Section III shows that solubilizing constituents of the conjugate will serve to achieve this goal, and such structural components as hydroxylated or carboxylated side groups, poly(ethylene oxide)-containing grafts or intrachain segments, as well as cationic or potentially cationing moieties in intrachain or extrachain attachment are outstanding examples of such solubility-enhancing constituents. Furthermore, an efficient mainchain flexibility will add a thermodynamic impetus to efficacious dissolution.

Yet another requirement for biological applications with which the early ferrocene polymer researchers were generally not concerned needs emphasis here: polymer backbone structures must be biodegradable to allow for slow excretion of the fragmented chain in the "spent" state. Nondegradable main-chain structures, as in vinyl polymers, are thus unacceptable. Successful synthetic approaches would typically have to provide for intrachain-type ester, amide, or urethane groups as potential backbone cleavage sites. Ideally, cleavage rates should be slow in order to avoid premature fragmentation while the conjugate still moves in central circulation. Finally, drug-anchoring groups should be incorporated that would complement any functional group attached to the ferrocene unit so as to provide polymer–ferrocene connecting links that are biofissionable for the ultimate release of monomeric ferrocene derivative. Such anchoring groups, as will be seen, are preferably of the carboxylic acid type, reacting with amino or hydroxyl groups on the metallocene, or of the primary or secondary amine type amenable to reaction with drug-attached carboxyl or carboxaldehyde functionality.

One of the most attractive polymer types to serve as drug carriers in accordance with the principles outlined in the foregoing is represented by the class of polyaspartamides. These nontoxic and nonimmunogenic polypeptides are obtainable from polysuccinimide by an aminolytic ring-opening reaction in anhydrous dimethylformamide.[50,51] The process leads to a backbone structure composed of randomly placed α- and β-peptidic units, these again appearing as D- and L-enantiomers. This composition has been found to be most beneficial insofar as in vivo degradation both by endo- and exopeptidases is severely retarded as β-peptidic and D forms provide a barrier to rapid, α-peptidase-mediated "unzipping" of the kind expected for a purely α-L-isomer structure and demonstrated for poly(L-glutamic acid) derivatives.[52] Backbone degradation in polyaspartamides is thus largely restricted to hydrolytic action. As an added bonus, the ring-opening reaction can be performed as a two-step or multistep process with two or more different amine nucleophiles used sequentially. Scheme 9 illustrates a typical two-step reaction in which the first amine provides a solubilizing group **S** and the second amine introduces the vital drug-binding functionality **F**. (Here and in subsequent representations, only the α-peptidic bonding pattern is shown for simplicity.) The α,β-DL-polyaspartamides were indeed, and continue to be, the "workhorse" carrier structures developed and used by the author's group on the strength of these features and the outstanding structural diversity inherent in the synthetic methodology permitting the construction of macromolecules in precisely predetermined compositions.[53]

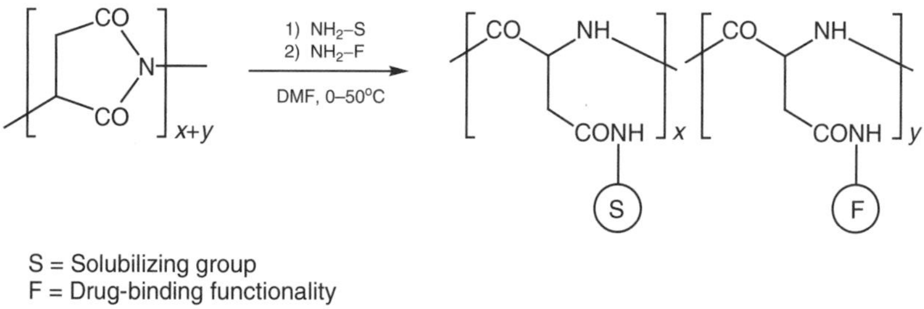

S = Solubilizing group
F = Drug-binding functionality

Scheme 9

B. Conjugates of Amide-Linked Ferrocene

Initial exploratory experiments demonstrated the practicability of binding the ferrocene unit to a variety of polyaspartamide structures of the general type shown in Scheme 9 with generation of water-soluble conjugates. The ferrocenylation agents, in the first place, were ferrocenylcarboxylic acids of the general structure $Fc-(CH_2)_n-COOH$ ($n=0-3$). Accordingly, the carriers were designed so as to possess primary or secondary amino groups as side chain terminals (**F** in Scheme 9), to which the ferrocenylation agents were coupled as the active N-succinimidyl esters.[38] Alternatively they were coupled directly as the free acids with mediation by the HBTU coupling agent (O-benzotriazolyl-N,N,N',N'-tetramethyluronium hexafluorophosphate).[54]

An exemplifying reaction sequence is depicted in Scheme 10, where the solubilizing group is represented by a tertiary amine functionality. Other carriers equipped with ethylenediamine or hydrazide side groups were allowed to react with carbonyl derivatives of ferrocene, typically ferrocenylcarboxaldehyde, giving conjugates containing imidazoline or hydrazone links.[38] On aqueous dialysis, the product polymers were isolated in the solid state by freeze drying of the retentate solutions. In that early work, coupling conditions were suboptimal, and the extent of ferrocenylation fell into the wide range of 25–90%. Exemplifying subunits with ferrocene attached via biocleavable amide links are depicted in Figure 2, which also displays two subunits representing product structures of α–keto derivatives of ferrocene.

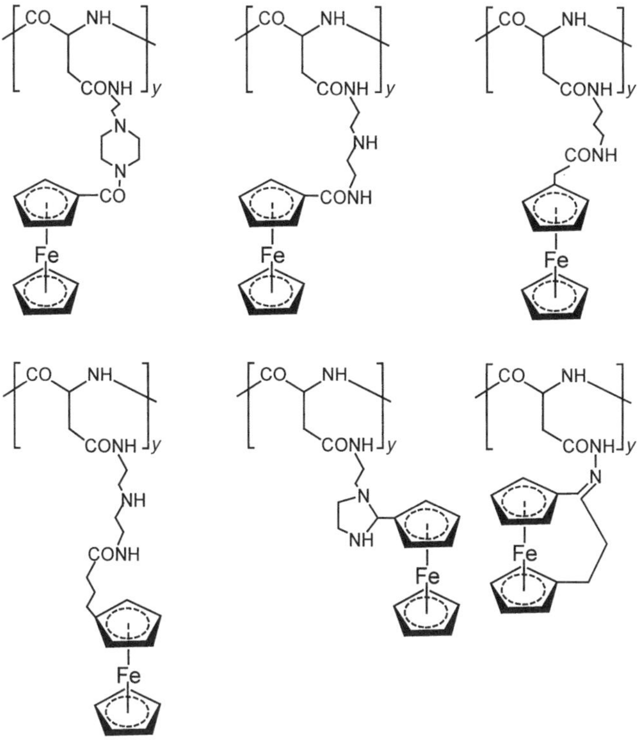

Figure 2 Exemplifying polyaspartamide–ferrocene subunits.

At this point the reader may wish to look briefly at the preferred analytical methods for the quantitative determination of the ferrocene contents in the reported conjugates, a task of crucial importance for their structural characterization and, hence, meaningful involvement in biomedical evaluation studies. Microanalytical determination of the iron contents will permit a calculation of the mass% ferrocene in the polymer and, thus, the extent of ferrocenylation, the latter defined as the mole fraction of ferrocene incorporated and expressed as a percentage of maximally available anchoring sites. In the author's laboratory, microanalytical assaying of the iron content, whether performed in house or by outside institutions, has proved to be afflicted with greater variability than spectroscopic techniques, such as proton magnetic resonance or electronic spectroscopy. These two spectroscopic methods, notably the convenient and time-saving ^{1}H NMR technique, have therefore been given preference. NMR spectra are preferentially recorded in D_2O at pD 10, thereby eliminating spurious N-protonation and occasionally observed hydrophobic interaction with resultant ferrocene proton signal attenuation. The ring protons of a ferrocene molecule substituted by electron donors (as in the butanoic acid derivative) give rise to a group of closely spaced resonance signals in the 4.2–4.0 ppm region. This part of the spectrum, relatively unencumbered by other resonances, represents a window that allows the ferrocene proton count to be conveniently correlated with the intensities of leading bands provided by the polymer; the extent of ferrocenylation may thus be directly read off the spectrum.

Before exploring the more recent conjugation studies discussed below, the reader will welcome some comments regarding the selection of the most suitable ferrocenylation agent. It had been shown in earlier electrochemical studies[55] that in the series of ferrocenylcarboxylic acids Fc–$(CH_2)_n$–COOH ($n=0$–3) the member with $n=3$,4-ferrocenylbutanoic acid, displays the lowest, that is, least positive formal reduction potential and, hence, the highest stability in its oxidized form. This observation was corroborated in the aforementioned project[54] and in a more recent study,[56] both involving conjugation of several ferrocenylcarboxylic acids, again, with an amine-functionalized polyaspartamide (Scheme 10).

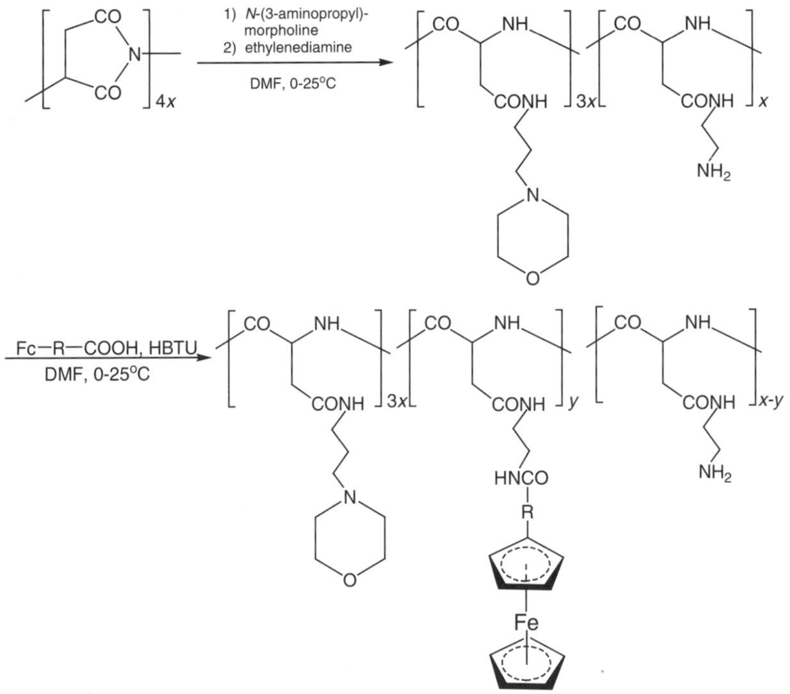

Scheme 10

Table 1 lists the $E^{\circ\prime}$ values, determined by cyclovoltammetry in aqueous ethanol for both the free and the polymer-bound acids, and the same trend toward lowest potential for the 4-ferrocenylbutanoic acid and its conjugate is apparent in both columns. Congruent with our working hypothesis (see Section II) that ferrocene exerts its antiproliferative activity as the oxidized, namely, cationic species, 4-ferrocenylbutanoic acid should thus be the preferred vehicle for the introduction of the metallocene into a polymeric structure, and this derivative, indeed, has become the ferrocenylation agent of choice in subsequent investigations.

Table 1 Formal Reduction Potentials of Selected Ferrocenylation Agents, Fc–R–COOH, and Derived Conjugates

Ferrocenylation Agent		$E^{\circ\prime}$ $(V)^a$	
Compound	R	Free Compound^b	Conjugate^c
Ferrocenoic acid	—	0.380	0.430
Ferrocenylacetic acid	$-CH_2-$	0.216	0.244
3-Ferrocenylpropanoic acid	$-(CH_2)_2-$	0.194	0.204
4-Ferrocenylbutanoic acid	$-(CH_2)_3-$	0.172	0.181
3-Ferrocenylbutanoic acid	$-CH(CH_3)CH_2-$	0.199	0.206
4-Ferrocenyl-4-hydroxypentanoic acid^d	$-C(CH_3)(OH)(CH_2)_2-$	0.193	0.214

^{a}Formal reduction potential, versus saturated calomel electrode, in aqueous ethanol. Data from Ref. 54. (See also Ref. 55.)
^{b}For comparison, $E^{\circ\prime}$ of ferrocene, 0.199 V.
^{c}Ferrocenylation agent carrier-bound per Scheme 10.
^{d}Na salt (free acid unstable).

The years following those earlier reports have witnessed further experimentation and preparative studies with the ultimate aim of providing conjugates of well-defined composition for biomedical evaluation of their cytotoxic effects on the cancerous cell. Again, polyaspartamides functionalized with primary amino groups have been the carriers of choice, and 4-ferrocenylbutanoic acid, in its free form or as its N-succinimidyl ester, has continued to serve as the ferrocenylation agent. However, in an effort to achieve complete substitution of the amine terminals provided by the carriers, it has proved necessary to choose more aggressive coupling conditions, in particular, a further increased excess of ferrocenylation agent. With 1.5 equiv of the agent for every equivalent of primary amino group the extent of metallocene incorporation is somewhat higher than in the early investigation,[38] attaining the 100% substitution level in roughly one-half of the trials. Optimal conversion, either by the HBTU-mediated direct coupling technique or by the active ester method involving the N-succinimidyl ester, is observed with a ferrocene/NH_2 ratio of 1.8. This modification[57–59] generally provides conjugates characterized by complete substitution of available amine anchoring sites. In those rare cases of undersubstitution still encountered in these polymer-homologous reactions, the resulting polymers can be retreated with ferrocenylation agent under duly adapted conditions, thereby converting to end products having the expected compositions.

A collection of leading contenders for bioassaying selected from these synthesized polyaspartamide-based conjugates is depicted in Schemes 11 and 12. In the former, we list conjugates containing *tert*-amine side groups as solubilizers, whereas the latter consists of conjugates solubilized by hydroxyl side groups. The listing contains a structure (**17**) characterized by the presence of a poly(ethylene oxide) segment R_2 introduced here for its solubilizing, protein-repelling, and other biomedically desired properties.

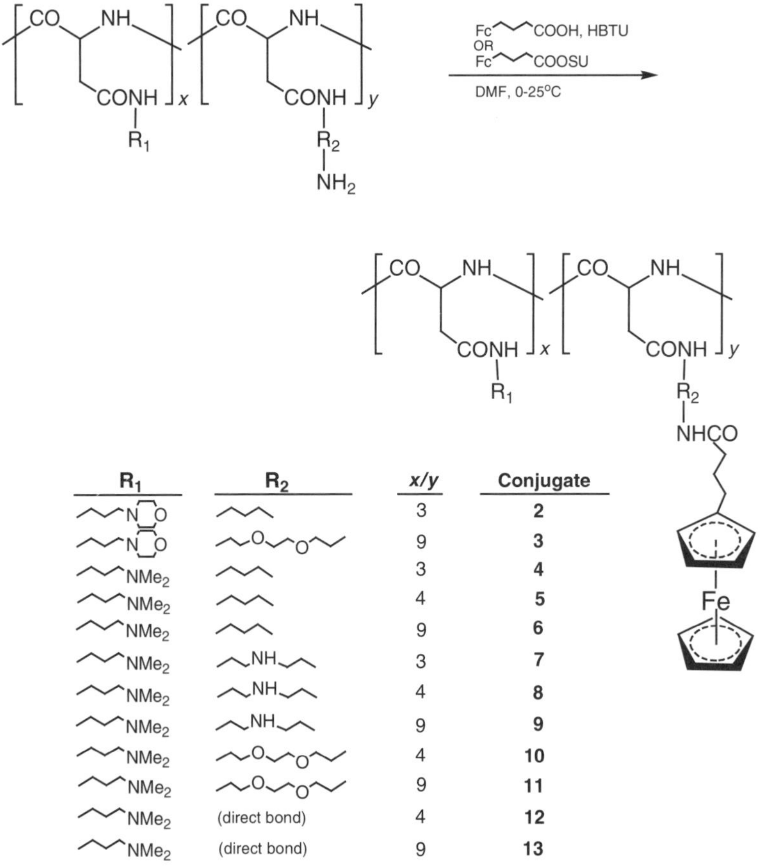

R_1	R_2	x/y	Conjugate
⌇⌇N‾O	⌇⌇	3	2
⌇⌇N‾O	⌇O⌇O⌇	9	3
⌇⌇NMe₂	⌇⌇	3	4
⌇⌇NMe₂	⌇⌇	4	5
⌇⌇NMe₂	⌇⌇	9	6
⌇⌇NMe₂	⌇NH⌇	3	7
⌇⌇NMe₂	⌇NH⌇	4	8
⌇⌇NMe₂	⌇NH⌇	9	9
⌇⌇NMe₂	⌇O⌇O⌇	4	10
⌇⌇NMe₂	⌇O⌇O⌇	9	11
⌇⌇NMe₂	(direct bond)	4	12
⌇⌇NMe₂	(direct bond)	9	13

Scheme 11

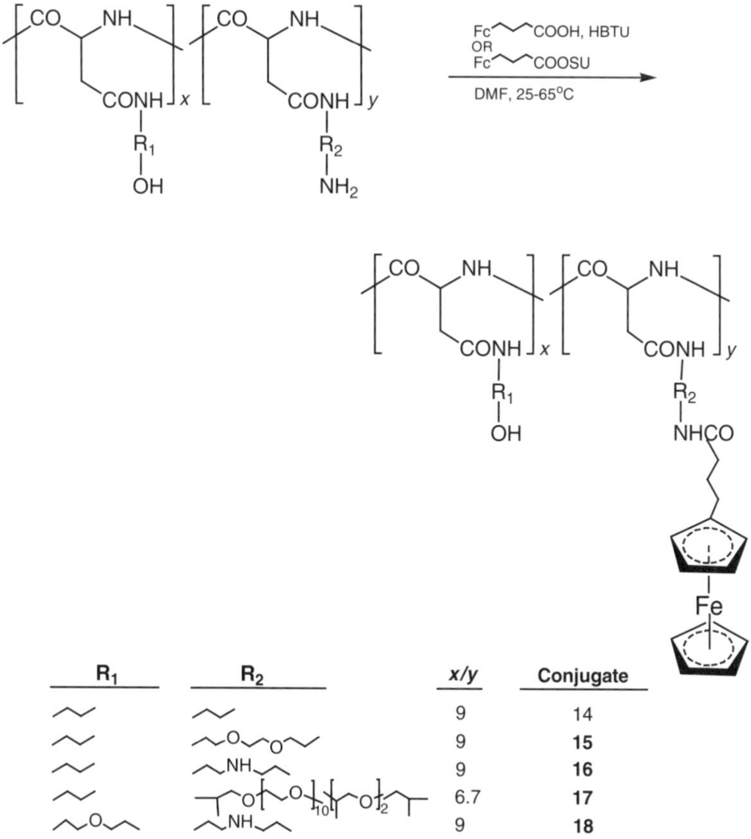

R_1	R_2	x/y	Conjugate
∿	∿	9	14
∿	∿O∿O∿	9	15
∿	∿NH∿	9	16
∿		6.7	17
∿O∿	∿NH∿	9	18

Scheme 12

A related conjugate (**19**) containing a methoxy-terminated poly(ethylene oxide) sidechain-grafted to an aspartamide subunit is shown in Scheme 13, where $x/y/w/z\text{-}w=6.42:0.67:1:0.25$. Here, apparently caused by the steric bulk of the graft, the extent of ferrocenylation has remained incomplete despite a ferrocene/NH_2 molar ratio increased to 2. All conjugates possess excellent solubility in aqueous media, with their upper solubility limits dictated solely by the rapid increase in solution viscosity. In contrast to the products of earlier studies, they are reported to remain soluble for a year or longer if stored at $-30°C$.

Whereas in the conjugates discussed so far ferrocene anchoring has involved primary amino groups attached to the carrier as side functionality, the metallocene may also be polymer-bound by acylation of secondary amine sites in the mainchain. This anchoring pattern will provide conjugates with the biofissionable >NCO– link in immediate vicinity of the backbone, a structural arrangement that may well have a bearing on the conjugates' biological behavior.[60] From a number of reported polymers falling into this category, we select just one illustrating a structural type represented by the polyamidoamine **20** of Scheme 14 ($n=3,21$). The carrier component, synthesized by a

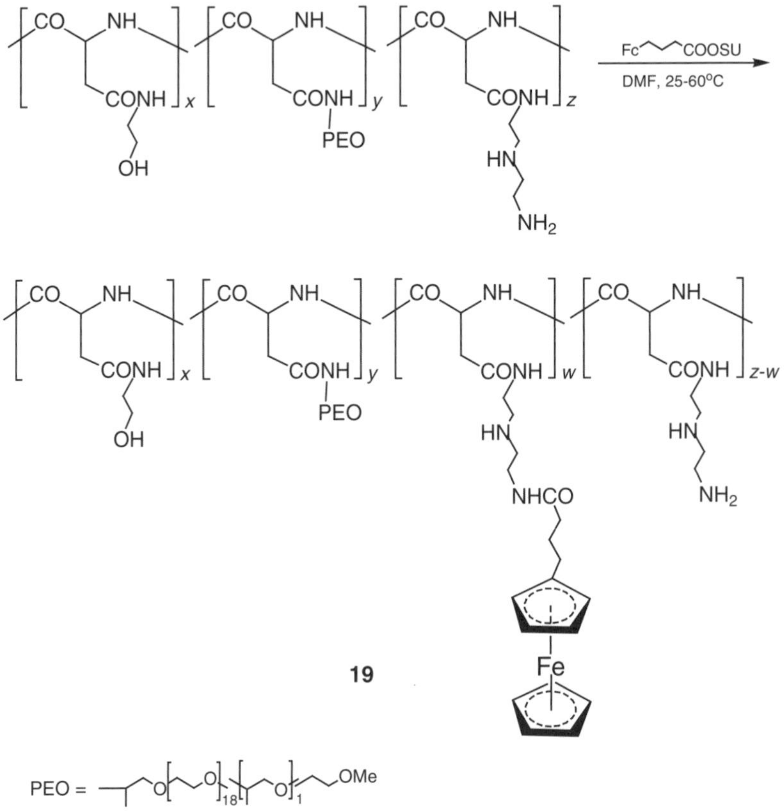

Scheme 13

Michael-type polyaddition process, possesses four secondary amine segments in the overall base recurring unit as shown, with the two subunits randomly distributed along the backbone. Treated with the active *N*-succinimidyl ester in dimethylformamide solution, it converts to a conjugate which, in principle, may thus contain four ferrocenyl

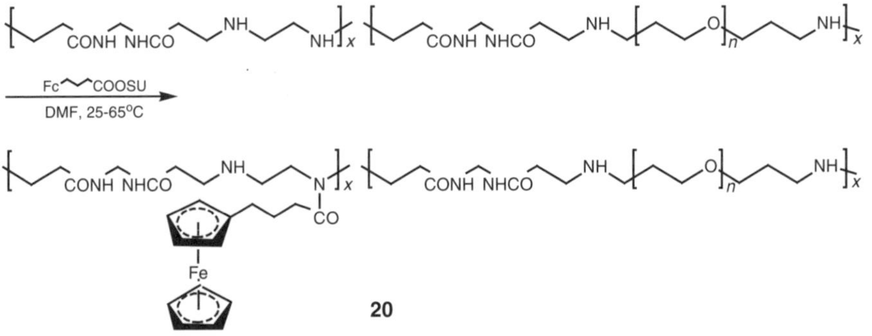

Scheme 14

groups in the base recurring unit. The accumulation of hydrophobic centers in a construct of this type would cause insolubility in water. Therefore, in the reported study the molar feed ratio of active ester to base recurring unit has been chosen (1.1 and 1.6, respectively) so as to favor the incorporation of just a single ferrocenyl group per unit as exemplified in **20**, with water solubility now duly retained.

C. Conjugates of Ester-Linked Ferrocene

The polymers covered in the preceding sections represent a class of conjugates in which ferrocene is carrier-bound through links of the amide type, which in the lysosomal compartment are cleaved enzymatically and, to a lesser extent, by hydrolysis. Replacement of the amide by an ester link will result in a connecting spacer that is, in the first place, amenable to hydrolytic fission in the mildly acidic lysosomal environment. Ferrocene polymers featuring ester-containing spacers will thus provide an interesting counterpart to amide-linked conjugates in biomedical screening studies. Contrasting with the latter class of conjugates, polymers containing ester-bound ferrocene have not attracted much attention, and published information is limited. The ester link is conveniently formed by allowing the free 4-ferrocenylbutanoic acid to couple, again in an aprotic solvent such as dimethylformamide, with carrier-attached hydroxyl groups, mediated by dicyclohexylcarbodiimide (DCC) and catalyzed by 4-(dimethylamino)pyridine (DMAP). With hydroxyl-functionalized polyaspartamide carriers, this approach, utilized in the author's laboratory,[61] has provided water-soluble conjugates **21–26** (Scheme 15), with R represented by an

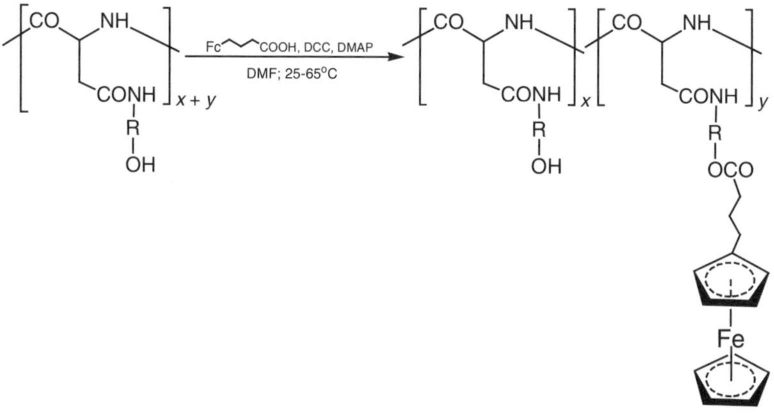

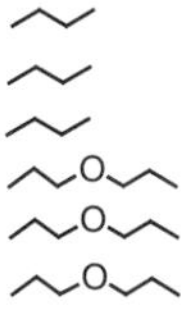

R	x/y	Conjugate
～～	9.8	21
～～	7.9	22
～～	4.3	23
～O～～	9.3	24
～O～～	7.2	25
～O～～	2.8	26

Scheme 15

ethylene or oxydiethylene segment. With molar ferrocene/hydroxyl feed ratios of 0.25–0.5, the ratio x/y, representing the number of unsubstituted over substituted base units, falls into the range of 3–10.

The same paper[61] describes a conjugate derived from a dihydroxyl-functionalized carrier prepared by an ester–amine polycondensation process from diethyl tartrate and 4,7,10-trioxa-1,13-tridecanediamine. With 1.8 mol of ferrocenylation agent used in the feed for every carrier base mole, the water-soluble product polymer **27** is obtained in a composition corresponding to $x/y=4$ (Scheme 16). For all conjugates here described the x/y ratios, determined by NMR spectroscopy, reflect ferrocene contents of approximately 10–25 mol%, a range considered convenient for biomedical applications.

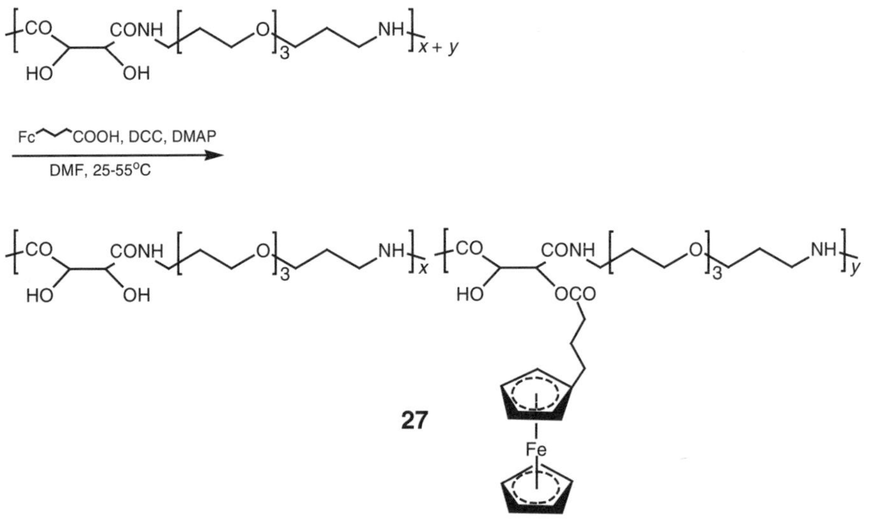

Scheme 16

V. BIOACTIVITY SCREENING

Having familiarized ourselves with the preparative methodology available for providing macromolecular ferrocene conjugates, we are ready now to explore their biomedical features. The fundamental issue then is whether we can expect the inherently inactive metallocene, on bioreversible conjugation with water-soluble carrier polymers, to convert to a potent cancer cell-killing agent, and further, whether such cytotoxic activity will be evident even though the conjugated metallocene in its administered form is in the Fe(II) (ferrocene) rather than Fe(III) (ferricenium) oxidation state.

Let us look at the available evidence. To this date, only in vitro screens assaying the antiproliferative activity of a limited number of selected conjugates have been reported, and most of these polymers are of the types depicted in

Schemes 11–14 characterized by amide-linked ferrocene. Test results obtained in murine cancer screens cannot reliably be correlated with drug behavior against mammalian, specifically human, cancers. Accordingly, with the exception of one project,[56] the presently discussed screens have been performed against human cancer cells grown in culture. The cancer models used include HeLa, a cervical epitheloid carcinoma line commonly considered sensitive to chemotherapeutic agents and frequently utilized as a standard, and the LNCaP human metastatic prostate adenocarcinoma, a cell line of special interest in view of the unsatisfactory therapeutic effectiveness of most anticancer drugs against secondary (i.e., metastatic) neoplasias. Finally included is Colo 320 DM, a human colon adenocarcinoma line. Intestinal cancers, in general, are poorly responsive to chemotherapy, and the Colo cell line as a typical colorectal cancer is known to possess some intrinsic drug resistance. The provisional results of tests against both HeLa and LNCaP, performed in two investigations on the polyaspartamide conjugates **2, 3, 4, 7,** and **11** (Scheme 11), have been published.[58,59] For any given polymer the findings reveal surprisingly similar activities against the two cell lines, showing that the metastatic LNCaP line features the same drug sensitivity as demonstrated by the HeLa standard. For example, IC_{50} values against the two lines are reported to be, respectively, 7.2 and 5 µg Fe/mL for conjugate **2,** 2.1 and 3 µg Fe/mL for conjugate **4,** and 2.3 and 2 µg Fe/mL for **11.**

The two studies also show that the incorporation of oligo(ethylene oxide) chains, either as a spacer component (**17;** Scheme 12) or as a sidechain graft (**19**), results in activities ($IC_{50} \geq 35$ µg Fe/mL) some 10-fold lower than observed for the remaining three conjugates devoid of oligo(ethylene oxide) constituents. The steric bulk of these chains, combined with their known protein-repelling properties, apparently retards, but does not preclude, access by the lysosomal proteolytic enzymes contributing to the drug release step. These structural peculiarities in **17** and **19,** while causing reduced in vitro efficaciousness, may well turn out to be advantageous for conjugate survival in central circulation on in vivo administration, and this topic needs further investigation.

The results of two follow-on studies performed on a broader scale are now on record,[62,63] and these are summarized in Table 2 in terms of IC_{50} values. In addition to HeLa and LNCaP, the screens include the Colo 320 DM line.

Excellent performance irrespective of the cell line used, comparing favorably with cisplatin data included for juxtaposition, is immediately apparent from the tabulation for the polymers **4–6, 8, 9, 11, 13,** and **20** distinguished by the presence of secondary or tertiary amino groups in their structures, with the majority of IC_{50} values below 1 µg Fe/mL. Somewhat lower activities are indicated for **15, 16,** and **18.** As a group, these last-named conjugates are characterized by hydroxyl terminals in the solubilizing groups, while devoid of amine functionality. Being unable to convert to protonated (and, thus, cationic) forms, they may experience increased resistance to pinocytotic cell entry (see Section III) and thus provide diminished drug bioavailability. (The exceptionally low IC_{50} value of 0.09 µg Fe/mL in the LNCaP column determined for **15** is inconsistent with other data and needs experimental confirmation). Whether the behavior of these hydroxyl-modified polymers will manifest itself in inferior performance under in vivo conditions remains an open, yet challenging question.

Table 2 Cell Culture Test Data for Selected Ferrocene Conjugates

	$IC_{50}{}^{b}$ (μg metal^{c}/mL)		
Conjugate^{a}	HeLa	Colo 320 DM	LNCaP
4			0.94
5	0.24	0.22	
6	0.49	0.42	0.27
8	0.32	2.18^{d}	
9	0.44	0.77	0.34
11	0.18	0.17	0.04
13	0.49	0.55	
15	19.5^{d}	4.06	0.09^{d}
16	2.27	0.21	4.57
18	7.87	1.56	
20	0.39	0.32	
Cisplatin	0.51	1.00	0.15

^{a} See Schemes 11 and 12.
^{b} Drug concentration required to achieve 50% cell growth inhibition, relative to drug-free control.
^{c} Fe for conjugates; Pt for cisplatin.
^{d} Excessive deviation from expected value; redetermination required.
Source: Data from Refs. 62 and 63.

Comparing now, for any given polymer, the data tabulated in the two parallel columns depicting IC_{50} values against the HeLa and LNCaP cell lines, one is led to confirm the earlier finding of equal activity against the two lines. More significantly still, a comparison of the data in the HeLa and Colo columns shows that, on average, activities against the "resistant" Colo cells are as high as, or even higher than, against the "sensitive" HeLa standard, a trend opposite to that for cisplatin. Figure 3 shows two representative curves depicting cell growth of LNCaP, in percent of drug-free control, as a function of conjugate concentration, in μg Fe/mL.

Whereas the plot drops off abruptly near 0.1 μg Fe/mL for conjugate **6**, it displays a more gradual descent right from the start for **11**, showing a plateau at about 35% cell growth before falling off further. Plateaus of this kind, in rare incidences even followed by a minor ascent with increasing concentration, have been observed occasionally with polyaspartamide–ferrocene conjugates, and this phenomenon needs further exploration.

At this point, then, our response to the question posed at the beginning of this section will be clearly affirmative: polymer conjugation of the inherently inactive ferrocene indeed represents an efficacious means to achieve significant cytotoxic activity, at least in in vitro systems, and this effect manifests itself even with the

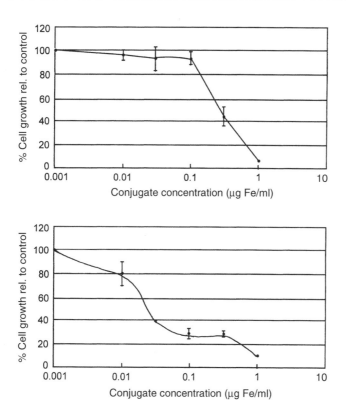

Figure 3 Representative plots of LNCaP cell growth, in % of drug-free control versus conjugate concentration, in μg Fe/mL, for **6** (top) and **11** (bottom). (Retraced from Ref. 63, with permission.)

metallocene polymer-bound and administered in its lower [Fe(II)] oxidation state. Metabolic activation of the metallocene through oxidation to the ferricenium state now remains a focal point of future investigations.

VI. SUMMARY AND CONCLUSIONS

Ferrocene, di-(η^5-cyclopentadienyl)iron(II), a key representative of the class of organometallic compounds known as *metallocenes*, has found numerous divergent applications in science and technology. The compound owes this multivaried behavior pattern largely to its unusual electronic structure and resultant oxidation–reduction properties. Attracted by the challenges offered in the realm of the biosciences, a major stream of investigative activities, as shown in Section II of this chapter, has flowed into biochemistry, where evidence of significant interaction of the nonpolymeric ferrocene–ferricenium system with the biological environment has been provided. In more recent years, the biomedical target presented by the proliferative cancer cell has emerged as a particularly rewarding topic in metallocene research. We have seen that strong antiproliferative, specifically, cell-growth-inhibiting action can be observed as cultured cancer cells are exposed in vitro to selected water-soluble ferrocene derivatives, especially those comprising the metallocene in its Fe(III) (ferricenium) state. We have also seen that excellent antineoplastic activity is evident even in certain in vivo systems.

Section III introduced the concept of polymer–drug conjugation, setting the stage for application of this innovative approach to the development of macromolecular ferrocene compounds for the purpose of further enhancing the metallocene's therapeutic effectiveness. Such macromolecular derivatives, possessing the crucial feature of water solubility, have now been synthesized from specially designed carrier polymers with the aid of tailored anchoring techniques, and they contain the ferrocene unit as an amide- or ester-bound side-group entity introduced by 4-ferrocenylbutanoic acid as the ferrocenylation agent, favored in most of the work on electrochemical grounds.

Cell culture tests screening selected conjugates for antiproliferative activity against several human cancer cell lines, although as yet quite preliminary, have demonstrated remarkable, and in many instances outstanding, performance. Of particular interest are the excellent results obtained against the Colo 320 DM and LNCaP lines, which are representative of moderately refractory, that is, drug-resistant cancers.

Where do we go from here now? Several alleys of research offer themselves for future pursuit in an effort to take the best possible advantage arising from the available research findings and promote further development of this class of organometallic polymers.

Restricting our consideration to immediate priorities, we foresee extended synthetic and cell culture-screening (i.e., in vitro) work to be conducted in an effort to identify for future in vivo studies the structural conjugate types emerging as leaders over the widest possible horizon. This will require a major synthesis program to provide a large variety of differently structured carriers and conjugates, and, herewith in conjunction, a biomedical evaluation effort to assess their antiproliferative properties against a broadened scope of human cell cultures including drug-sensitive cell lines and their resistant counterparts. Forth-coming investigations should focus on variations of these key structural elements: (1) *mainchain* [length, biodegradation, effect of flexibilizing segments such as oligo(ethylene oxide) units]; (2) *solubilizing and lysosomotropic groups* (type, frequency, and basicity; intra- or extrachain location); (3) *drug-binding systems* (spacer type, frequency, length and flexibility, effect on reduction potential of attached ferrocenyl group; type of biofissionable group and susceptibility to pH-dependent hydrolytic and enzymatic attack).

Additional topics of preeminent interest to be addressed include combination drug effects (multidrug conjugation to a single carrier), photosensitizing effects, preoxidation to ferricenium conjugates and comparative studies with the respective ferrocene analogs, long-term conjugate storage stability, and extended in vitro toxicity determinations.

Further down the trek, still within the precinct of in vitro methodology, are studies of drug release kinetics, hemolytic properties, and of the mechanism(s) of cytotoxic activity, paving the way for ultimate in vivo assessment of therapeutic effectiveness. With the low systemic toxicity of ferrocene conjugates[64] taken into account it is apparent from the performance data so far obtained that macromolecular ferrocene conjugates will lend themselves as excellent candidates for further research in cancer chemotherapy, notably in the treatment of drug-resistant malignancies.

VII. ACKNOWLEDGMENTS

This ongoing research program has been sustained by grants from Council of this University, the Anglo American Chairman's Fund, and the Cancer Association of South Africa in conjunction with the THRIP Project. The author gratefully acknowledges the dedicated contributions by his coworkers, here and abroad, named in the references. The author is indebted also to Kitty Banda and Pat Cawthorn for their excellent clerical and artwork assistance, and to Iain Burns and his team in the Research Office for their loyal and unceasing administrative support.

VIII. REFERENCES

1. M. Rosenblum, *Chemistry of the Iron Group Metallocenes*, Wiley, New York, 1965.

2. A Togni, T. Hayashi, eds., *Ferrocenes*, VCH, Weinheim, 1995.

3. E. W. Neuse, *Macromol. Sci. Chem.* **A16**, 3 (1981) (a review of polymetallocenylenes).

4. T. J. Kealy, P. L. Pauson, *Nature* **168**, 1039 (1951).

5. E. W. Neuse, in *Metal Containing Polymer Systems*, J. E. Sheats et al., eds., Plenum Press, New York, 1985, p. 99.

6. Selected references of unclassified publications cited in E. W. Neuse, J. R. Woodhouse, G. Montaudo, C. Puglisi, *Appl. Organomet. Chem.* **2**, 53 (1988).

7. K. E. Dombrowski, W. Baldwin, J. E. Sheats, *J. Organomet. Chem.* **302**, 281 (1986).

8. B. Halliwell, J.M.C. Gutteridge, *Free Radicals in Biology and Medicine*, 2nd ed., Clarendon Press, Oxford, 1990.

9. R. Epton, M.E. Hobson, G. Marr, *J. Organomet. Chem.* **34**, C23 (1977).

10. W. A. Kornicker, B. L. Vallee, *Ann. NY Acad. Sci.* **153**, 689 (1968).

11. M. J. Carney, J. S. Lesniak, M. D. Likar, J. R. Pladziewicz, *J. Am. Chem. Soc.* **106**, 2565 (1984).

12. J. R. Pladziewicz, M. S. Brenner, D. A. Rodeberg, M. D. Likar, *Inorg. Chem.* **24**, 1450 (1985).

13. B. W. Carlson, L. L. Miller, P. Neta, J. Grodkowski, *J. Am. Chem. Soc.* **106**, 7233 (1984); see also B. W. Carlson, L.L. Miller, *J. Am. Chem. Soc.* **105**, 7453 (1983).

14. M. S. McDowell, J. H. Espenson, A. Bakač, *Inorg. Chem.* **23**, 2232 (1984).

15. S. R. Logan, *Z. Physikal. Chem.* **167**, 41 (1990).

16. S. R. Logan, G. A. Salmon, *J. Chem. Soc., Perkin Trans. II* 1781 (1983).

17. S. R. Logan, *J. Chem. Soc., Perkin Trans. II* 751 (1989).

18. A. M. Joy, D. M. L. Goodgame, I. J. Stratford, *Int. J. Radiation Oncol. Biol. Phys.* **16**, 1053 (1989).

19. D. Osella, M. Ferrali, P. Zanello, F. Laschi, M. Fontani, C. Nervi, G. Cavigiolio, *Inorg. Chim. Acta* **306**, 42 (2000).

20. W. Troll, G. Witz, B. Goldstein, D. Stone, T. Sugimura, in *Carcinogenesis*, Vol. 7, E. Hecker et. al., eds., Raven Press, New York, 1982, p. 593.

21. T. W. Kensler, D. M. Bush, W. J. Kozumbo, *Science* **221**, 75 (1983).

22. P. J. O'Brien, *Env. Health Persp.* **64**, 219 (1985).

23. G. Powis, *Free Radical Biol. Med.* **6**, 63 (1989).

24. P. Kovacic, W. J. Popp, J. R. Ames, M. D. Ryan, *Anti-Cancer Drug Design* **3**, 205 (1988).

25. T. W. Kensler, M. A. Trush, *Biochem. Pharmacol.* **32**, 3485 (1983).

26. M. D. Sevilla, P. Neta, L. J. Marnett, *Biochem. Biophys. Res. Commun.* **115**, 800 (1983).

27. P. Köpf-Maier, H. Köpf, E. W. Neuse, *J. Cancer Res. Clin. Oncol.* **108**, 336 (1984); see also *Angew. Chem. Int. Ed.* **23**, 456 (1984).

28. P. Köpf-Maier, H. Köpf, E.W. Neuse, Eur. Patent 0153636 B1 (Dec. 6 1989).

29. M. Wenzel, Y. Wu, E. Liss, E. W. Neuse, *Z. Naturforsch.* **43c**, 963 (1988).

30. E. W. Neuse, F. Kanzawa, *Appl. Organometal. Chem.* **4**, 19 (1990).

31. P. Köpf-Maier, H. Köpf, *Struct. Bond.* **70**, 103 (1988).

32. R. W. Mason, K. McGrouther, P. R. R. Ranatunge-Bandarage, B. H. Robinson, J. Simpson, *Appl. Organometal. Chem.* **13**, 163 (1999).

33. A. Houlton, R. M. G. Roberts, J. Silver, *J. Organomet. Chem.* **418**, 107 (1991).

34. L. V. Popova, V. N. Babin, Yu. A. Belousov, Yu. S. Nekrasov, A. E. Snegireva, N. P. Borodina, G. M. Shaposhnikova, O. B. Bychenko, P. M. Raevskii, N. B. Morozova, A. I. Ilyina, K. G. Shchitkov, *Appl. Organomet. Chem.* **7**, 85 (1993).

35. V. N. Babin, P. M. Raevskii, K. G. Shchitkov, L. V. Snegur, Yu. S. Nekrasov, *Mendeleev Chem. J.* **39**, 17 (1995).

36. C. G. Hartinger, A. A. Nazarov, V. B. Arion, G. Giester, M. Jakupec, M. Galanski, B.K. Keppler, *New J. Chem.* **26**, 671 (2002).

37. E. W. Neuse, C. W. N. Mbonyana, *Int. Symp. Inorg. Organometal. Polym.*, Miami Beach, Florida, Sept. 8–9, 1989; *Abstr. Polym. Mater. Sci. Eng.* 20.

38. E. W. Neuse, C. W. N. Mbonyana, in *Inorganic and Metal-Containing Polymeric Materials*, J. Sheats et al., eds., Plenum Press, New York, 1990, p. 139.

39. H. Maeda, *Adv. Drug Delivery Rev.* **6**, 181 (1991); reviewed by H. Maeda, J. Wu, T. Sawa, Y. Matsumura, K. Hori, *J. Contr. Release* **65**, 271 (2000).

40. H. Ringsdorf, *J. Polym. Sci. Polym. Symp.* **51**, 135 (1975).

41. R. Duncan, J. Kopeček, *Adv. Polym. Sci.* **57**, 51 (1984), Section 3.2 (a review).

42. H. J. -P. Ryser, W. -C. Shen, N. Morad, *ACS Polym. Preprints* **27**, 15 (1986).

43. D. Putnam, J. Kopeček, *Adv. Polym. Sci.* **122**, 55 (1995).

44. S. E. Matthews, C. W. Pouton, M. D. Threadgill, *Adv. Drug Delivery Rev.* **18**, 219 (1996).

45. C. J. T. Hoes, J. Feijen, in *Drug Carrier Systems*, F. H. D. Roerdink, A. M. Kroon, eds., Wiley, London, 1989, p. 57.

46. E. W. Neuse, "Metallocene Polymers," *Encyl. Polym. Sci. Technol.* **8**, 667 (1968).

47. E. W. Neuse, H. Rosenberg, *J. Macromol. Sci. Rev. Macromol. Chem.* **C4**, 1 (1970).

48. E. W. Neuse, E. Quo, *Bull. Chem. Soc. Jpn.* **39**, 1508 (1966).

49. E. W. Neuse, *Ferrocene Polymers*, U.S. Patent 3,238,185 (March 1, 1966).

50. P. Neri, G. Antoni, *Macromol. Synth.* **8**, 25 (1982). See also: P. Neri, G. Antoni, F. Benvenuti, F. Cocola, G. Gazzei, *J. Med. Chem.* **16**, 893 (1973).

51. J. Drobnik, V. Saudek, J. Vlasák, J. Kalal, *J. Polym. Sci. Polym. Symp.* **66,** 65 (1979).

52. J. Pytela, R. Kotva, F. Rypáček, *J. Bioact. Compat. Polym.* **13,** 198 (1998).

53. E. W. Neuse, A. G. Perlwitz, A. P. Barbosa, *J. Appl. Polym. Sci.* **54,** 57 (1994), and preceding papers in this series.

54. J. C. Swarts, E. W. Neuse, G. J. Lamprecht, *J. Inorg. Organomet. Polym.* **4,** 143 (1994).

55. N. F. Blom, E. W. Neuse, H. G. Thomas, *Transition Met. Chem.* **12,** 301 (1987).

56. J. C. Swarts, D. M. Swarts, D. M. Maree, E. W. Neuse, C. La Madeleine, J. E. Van Lier, *Anticancer Res.* **21,** 2033 (2001).

57. M. G. Meirim, E. W. Neuse, G. A. Caldwell, *J. Inorg. Organomet. Polym.* **7,** 71 (1997).

58. G. Caldwell, M. G. Meirim, E. W. Neuse, C. E. J. van Rensburg, *Appl. Organomet. Chem.* **12,** 793 (1998).

59. G. Caldwell, M. G. Meirim, E. W. Neuse, K. Beloussow, W. -C. Shen, *J. Inorg. Organomet. Polym.* **10,** 93 (2000).

60. M. G. Meirim, E. W. Neuse, G. Caldwell, *J. Inorg. Organomet. Polym.* **8,** 225 (1998).

61. E. W. Neuse, M. G. Meirim, D. D. N'Da, G. Caldwell, *J. Inorg. Organomet. Polym.* **9,** 221 (1999).

62. M. T. Johnson, E. Kreft, D. D. N'Da, E. W. Neuse, C. E. J. van Rensburg, manuscript submitted.

63. Unpublished work with C. -J. Lim and W. -C. Shen; see also C. -J. Lim, M.Sc. thesis, Univ. Southern California, Los Angeles, CA, Aug. 2002.

64. B. Schechter, G. Caldwell, E. W. Neuse, *J. Inorg. Organomet. Polym.* **10,** 177 (2000).

CHAPTER 7

Polymeric Platinum-Containing Drugs in the Treatment of Cancer

Deborah W. Siegmann-Louda

Florida Atlantic University, Boca Raton, Florida

Charles E. Carraher Jr.

Florida Atlantic University, Boca Raton, Florida and Florida Center for Environmental Studies, Palm Beach Gardens, Florida

CONTENTS

I. INTRODUCTION   120

II. BASIC MECHANISMS OF Pt(II) COMPLEX FORMATION   121

III. NOMENCLATURE   125

IV. CURRENTLY APPROVED PLATINUM-CONTAINING COMPOUNDS   125

V. PROPERTIES OF CISPLATIN   127

VI. STRUCTURE–ACTIVITY RELATIONSHIPS   130

VII. POLYMER–DRUG CONJUGATION STRATEGY AND POSSIBLE BENEFITS   133
    A. Polymers as Carriers   134
    B. Polymers as Drugs   135
    C. General   136

Macromolecules Containing Metal and Metal-Like Elements,
Volume 3: Biomedical Applications, edited by Alaa S. Abd-El-Aziz,
Charles E. Carraher Jr., Charles U. Pittman Jr., John E. Sheats, and Martel Zeldin
ISBN: 0-471-66737-4   Copyright © 2004 John Wiley & Sons, Inc.

VIII. MAINCHAIN-INCORPORATED *cis*-DIAMINE-
        COORDINATED PLATINUM   137
        A. Simple Amine Derivatives   137
        B. Amino Acid Derivatives   141
        C. Other Nitrogen–Platinum Products   143
        D. Solution Stability   144
        E. Thermal Stability   145
        F. Antiviral Activity   145

  IX. PLATINUM CARRIER-BOUND COMPLEXES VIA
        NITROGEN DONOR LIGANDS   147
        A. Pt-Polyphosphazenes   147
        B. Slowly Biofissionable Pt–N Complexes Anchored through
            Primary and Secondary Amines   149
        C. Biofissionable Pt–N Complexes Anchored through
            Primary and Secondary Amines   154

   X. Pt–O-BOUND POLYMERS   161

  XI. MIXED Pt–O/Pt–N-BOUND POLYMERS   180

 XII. FUTURE WORK   182

XIII. ACKNOWLEDGMENTS   184

XIV. REFERENCES   185

I. INTRODUCTION

Malignant neoplasms, cancers, are the second leading cause of death in the United States. In 1964 Rosenberg and coworkers discovered that in the presence of platinum electrodes *Escherichia coli* failed to divide but continued growing, giving filamentous cells.[1] The cause of the inhibition of cell division was found to be cis-diaminedichloroplatinum (II), also known as *cis*-DDP and cisplatin.

cis-Diaminedichloroplatinum(II), cisplatin, has been known since 1845 as Peyrone's salt, and much research has been based on cisplatin itself as well as monomeric and polymeric derivatives because of the ability of cisplatin to inhibit a number of cancers.

In 1972 clinical trials using cisplatin were initiated. The use of cisplatin has been complicated because of its toxicity, bringing about a number of negative side effects, including gastrointestinal, auditory, hematopoietic, immunosuppressive, and renal dysfunction.[2–4]

Cisplatin is the most widely used general anticancer agent, most often employed along with at least one other anticancer drug that operates through another mechanism.[5–11] It shows especially good results with taxol, 5-fluorouracil, methotrexate, bleomycin, and cytarabine. These combinations are active against refractive diseases including melanoma and breast cancer.[8-11] While used in a wide variety of cancers, it is the drug of choice for certain cancers. It is 70–90% effective against testicular cancer, is highly effective against ovarian cancer, and is widely used for neck and head cancers.[7,12,13]

Expenditures for cisplatin are in the range of $500 million per year. [14]

Cisplatin is normally administered by the intravenous route and less frequently delivered via the intraperitoneal route.

It is the purpose of the present efforts to incorporate the platinum atom in various polymers to produce *cis*-platin-related drugs with enhanced anticancer activity and overall lowered toxicity.

II. BASIC MECHANISMS OF Pt(II) COMPLEX FORMATION

Transition metal square–planar complexes generally contain eight d electrons and are almost always diamagnetic. This includes complexes of Pt^{2+}, Pd^{2+}, Au^{3+}, Rh^{1+}, and Ir^{1+}. While such complexes can undergo other reactions such as redox processes, we shall focus on substitution reactions. Good reviews of square–planar substitution reactions are available.[15-17] The following is a summary of some of these substitution processes, with emphasis on those involved with polymer formation. These substitution reactions are the most widely studied of the transition metal square–planar complex reactions.

As with many reactions, the final structure depends on the starting material and the order of addition, as illustrated with the following simple reactions.[18]

$$
\begin{array}{ccc}
\text{Cl} & & \text{Cl} \\
| & & | \\
[\text{Cl-Pt-Cl}]^{2-} + NH_3 \rightarrow & [\text{Cl-Pt-NH}_3]^{1-} & \quad (1)\\
| & & | \\
\text{Cl} & & \text{Cl}
\end{array}
$$

$$
\begin{array}{ccc}
\text{Cl} & & \text{NO}_2 \\
| & & | \\
[\text{Cl-Pt-NH}_3]^{1-} + NO_2^{1-} \rightarrow & [\text{Cl-Pt-NH}_3]^{1-} & \quad (2)\\
| & & | \\
\text{Cl} & & \text{Cl}
\end{array}
$$

$$
\begin{array}{ccc}
\text{Cl} & & \text{Cl} \\
| & & | \\
[\text{Cl-Pt-Cl}]^{2-} + NO_2^{1-} \rightarrow & [\text{Cl-Pt-NO}_2]^{2-} & \quad (3)\\
| & & | \\
\text{Cl} & & \text{Cl}
\end{array}
$$

$$[Cl\text{-}\overset{\displaystyle Cl}{\underset{\displaystyle Cl}{\overset{|}{\underset{|}{Pt}}}}\text{-}NO_2]^{2-} + NH_3 \rightarrow [H_3N\text{-}\overset{\displaystyle Cl}{\underset{\displaystyle Cl}{\overset{|}{\underset{|}{Pt}}}}\text{-}NO_2]^{1-} \qquad (4)$$

$$[Cl\text{-}\overset{\displaystyle Cl}{\underset{\displaystyle Cl}{\overset{|}{\underset{|}{Pt}}}}\text{-}Cl]^{2-} + 2PR_3 \rightarrow R_3P\text{-}\overset{\displaystyle Cl}{\underset{\displaystyle Cl}{\overset{|}{\underset{|}{Pt}}}}\text{-}PR_3 \qquad (5)$$

$$[R_3P\text{-}\overset{\displaystyle PR_3}{\underset{\displaystyle PR_3}{\overset{|}{\underset{|}{Pt}}}}\text{-}PR_3]^{2+} + 2Cl^{1-} \rightarrow R_3P\text{-}\overset{\displaystyle Cl}{\underset{\displaystyle Cl}{\overset{|}{\underset{|}{Pt}}}}\text{-}PR_3 \qquad (6)$$

$$[Cl\text{-}\overset{\displaystyle Cl}{\underset{\displaystyle Cl}{\overset{|}{\underset{|}{Pt}}}}\text{-}Cl]^{2-} + 2NH_3 \rightarrow Cl\text{-}\overset{\displaystyle NH_3}{\underset{\displaystyle Cl}{\overset{|}{\underset{|}{Pt}}}}\text{-}NH_3 \qquad (7)$$

$$[H_3N\text{-}\overset{\displaystyle NH_3}{\underset{\displaystyle NH_3}{\overset{|}{\underset{|}{Pt}}}}\text{-}NH_3]^{2+} + 2Cl^{1-} \rightarrow H_3N\text{-}\overset{\displaystyle Cl}{\underset{\displaystyle Cl}{\overset{|}{\underset{|}{Pt}}}}\text{-}NH_3 \qquad (8)$$

These reactions show a dependence on the concentration of the incoming ligand.[19,20] Generally, under pseudo-first-order conditions, plots of the log of the metal complex concentration versus time are linear with a slope of $k_{observed}$.[15–17,19,20] Varying the concentration and nature of the ligand gives a series of linear plots for each ligand so that the rate law has a ligand-independent term and a ligand-dependent term. A strong solvent effect is also found for such reactions. The ligand-dependent term is associated with a nucleophilic attack by the entering ligand. Such square–planar complexes have open sites above and below the square plane with orbitals available for attack. The ligand independent part of the reaction may be explained several ways. The most widely accepted explanation is that the second term involves a direct nucleophilic attack by the solvent on the square–planar complex.

The overriding factor determining the structure of Pt(II) complexes is the effect that the ligand trans to the substitution site has on the rate of substitution.[15–17] This effect is called the *trans effect* and was recognized in the early 1900s. The general trans effect order is[21]

CN^{1-}, CO, $C_2H_4 > H^{1-}$, PR_3, $> CH_3^{1-} > Ph^{1-}$, I^{1-}, $NO_2^{1-} > Cl^{1-}$, $Br^{1-} > NH_3$, Py, H_2O, OH^{1-}

This order spans a difference in about 10^6 in rate. This trend is important in determining the tendency to form, almost exclusively, the product predicted by the trans effect. It is also important in determining the stabilities of complexes as well as predicting which complex may be formed in the presence of competing ligands, such as in the blood where various water, chloride ion and other ligand species are present.

Before continuing we should distinguish between the terms *trans effect* and *trans influence*. The former describes the effect of the ligand on the rate of substitution, while the latter describes the effect of the trans ligand on ground-state properties.[15–17] Thus, the term *trans influence* concerns ground-state influences of a trans ligand. The *trans effect* describes influences of the trans ligand on both the ground and transition state.

The trans effect has been explained in a number of ways.[15–17,22–24] The best description involves both π-accepting and σ donating capabilities of the ligand. The trans ligands in a square–planar complex share a p orbital. Ligands that are strong σ donors contribute a greater electron density to the shared p orbital and weaken the leaving-group bond. The expected order for the trans effect from sigma donation is

$$H^{1-}, PR_3 > I^{1-}, CO, CH_3^{1-}, CN^{1-} > Br^{1-} > Cl^{1-} > NH_3 > OH^{1-}$$

which is similar to the general order given before, where CO and CN^{1-} are out of order.

Carbon monoxide and the cyanide ion are believed to greatly stabilize the transition state because they are π-accepting ligands. A ligand with π acceptance capability can remove electron density from the metal atom, stabilizing the transition state. The order of π-bonding ability is

$$CO, C_2H_4 > CN^{1-} > NO_2^{1-} > I^{1-} > Br^{1-} > Cl^{1-} > NH_3 > OH^{1-}$$

and is said to account for the lack of conformity when considering only σ donation. Thus, both σ and π contributions appear to be important in describing the trans effect.

While the trans effect is the major factor governing square–planar substitution reactions other factors contribute. These factors are the electronic effect of *cis* ligands (generally relatively unimportant), leaving-group effects, effect of the entering ligand, and solvent effects.

The effect of the leaving group has been studied using platinum diene systems as below.[25]

$$[Pt(dien)X]^{1+} \; + \; Py \; \longrightarrow \; [Pt(dien)Py]^{2+} \; + \; X^{1-} \tag{9}$$

The study indicates that there is a substantial amount of bond breakage in the transition state with the amount of bond breakage depending on the specific reaction studied. The order of leaving-group dependence is very similar to the inverse of the trans effect order. Thus, the trans effect depends on the strength of the bonding (a combination of σ and π) of the ligand, with the more strongly bonded ligands dissociating more slowly from the 5-coordinate intermediate.

There is a dependence on the entering nucleophile, as expected for nucleophilic attacks. The general order is[18,22,23,26,27]

$$PR_3 > N_3{}^{1-}, I^{1-} > NO_2{}^{1-} > Br^{1-} > Py > Cl^{1-}, NH_3 > H_2O > OH^{1-}$$

which is consistent with the polarizability or softness of the incoming nucleophile being important. The hard–soft theory says that soft ligands (nucleophiles) prefer soft substrates while hard ligands prefer hard sites. The observed trend is that the larger ligands are more effective nucleophiles for Pt^{2+}, consistent with Pt^{2+} being a soft site.

The reaction sited below was studied as a function of a number of entering groups, Y:[20,28,29]

$$\text{\textit{trans}-PtL}_2\text{Cl}_2 \quad + \quad Y \quad \rightarrow \quad \text{\textit{trans}-PtL}_2\text{ClY} \quad + \quad Cl^{1-} \qquad (10)$$

The order of entering ligand Y nucleophilicities on reaction rate is similar to the observed trans-effect ordering. This is believed to be because of the relative positions of the entering ligand Y and the trans group in the trigonal bipyramidal transition state. Here, the entering ligand and trans ligand occupy similar positions and so should influence the rate similarly. Thus, entering and leaving ligand effects are similar because they occupy approximately equivalent positions in the transition state.

As noted before, the rate law for square–planar substitution reactions includes a term that is believed associated with the solvent. Thus, it is expected that the reaction rates are solvent-dependent. The influence on reaction rate depends on the relative nucleophilicity of the solvent and the entering ligand. In highly coordinating solvents such as DMSO, water, ethanol, and propanol (which can occupy octahedral sites above and below the square plane), reaction occurs by a ligand-independent pathway. In solvents with low coordination ability such as carbon tetrachloride, benzene, and acetone, the reaction proceeds solely by attack of the ligand on the square–planar complex. In solvents that show good coordination ability the reaction rates show a direct dependence on the nucleophilicity of the solvent. Reaction rates are faster for those solvents that have poor coordinating ability, where the ligands are not solvated.[30,31]

The major effort in the synthesis of platinum-containing polymers involves reactions between the tetrahaloplatinate(II) and a nitrogen-containing site. Similar reactions occur with other Lewis bases, most resulting in the formation of the cis compound as the major product. As expected, nucleophilicity of the Lewis base is important in considering the relative rate of reaction. Thus, reaction with alcohols is slow in comparison to reaction with salts of carboxylic acids. Further, much effort has involved reactions employing other platinum-containing reactants. Products derived from reaction with alcohols and acid salts will also be discussed in this review, and they are also potentially important agents in the war against cancer.

It is important for those not familiar with Pt–N chemistry to note that it is customary to simply assume that bonding between the platinum and nitrogen atom is coordinate covalent bonding and to omit the placement of an arrow between the Pt and N, that is N–Pt rather than N—>Pt. It is also customary to omit the positive charge that is on the nitrogen, for instance R_2HN–Pt rather than R_2HN^+–Pt.

Analogous reactions employing palladium have also been studied for polymer synthesis [32–39] as well as monomer synthesis.[40–45] Some of these materials are potential anticancer drugs since they show inhibition of certain cancer cell lines.[33,35,37,39]

These Pd^{2+} materials react about 10^5 times faster than corresponding Pt^{2+} complexes. This is ascribed to the weaker Pd–L bond. Thus, a greater effective nuclear charge gives a stronger metal–ligand bond (for ligands that do not accept electron density from the metal site) as one goes from left to right in the periodic table for the same row, and as one goes from the first to second to third row. (Hard–soft considerations are also probably in play here.)

Substitution reactions of Pd^{2+} are similar to those of Pt^{2+} with respect to the trans ligand, leaving group, entering ligand, and solvent effect. The kinetics are similar with both solvent- and ligand-based terms, except the solvent term is more important for the palladium complexes.

III. NOMENCLATURE

Organometallic chemistry suffers from inconsistent naming of compounds. As in the case of organic chemistry, common names are often used and are more easily recognized than the official or IUPAC name.Thus, the central platinum compound, *cis*-diaminedichloroplatinum(II) is often found referred to as *cis*-diamine-dichloro-platinum (II), *cis*-diaminedichloroplatinum II, diaminedichloroplatinum II, and other variations on the same theme. Here, we will attempt to use the correct form. Further, *cis*-diaminedichloroplatinum(II) also has several common names such as cisplatin, CDDP, *cis*-DDP, and cisplatinum. Cisplatin is the most widely used of these names. Here, we will use both cisplatin and *cis*-DDP.

The platinum compounds are also alternately referred to as *platinum complexes*, *compounds*, and *conjugates*. Here we will use all of these designations.

Many authors choose to use a double "m" when referring to these diamine compounds, such as *cis*-diammminedichloroplatinum(II), but here we will use the single-"m" spelling.

IV. CURRENTLY APPROVED PLATINUM-
CONTAINING COMPOUNDS

The use of monomeric platinum drugs will now be introduced, providing a background for considering polymeric platinum species as bioactive agents.

While thousands of cisplatin analogs have been synthesized and screened only about 28 platinum compounds have entered clinical trials as anticancer agents.[7,46] Of these only four are currently approved. Those approved are cisplatin, carboplatin, oxaliplatin, and nedaplatin. Only the first two are commercially available for general use in the treatment of cancer. The structures are given below **(1–4)**.

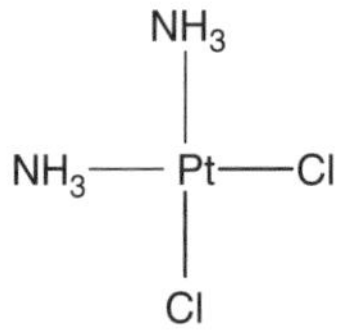

1

[cisplatin, *cis*-DDP:
cis-diaminedichloroplatinum(II)]

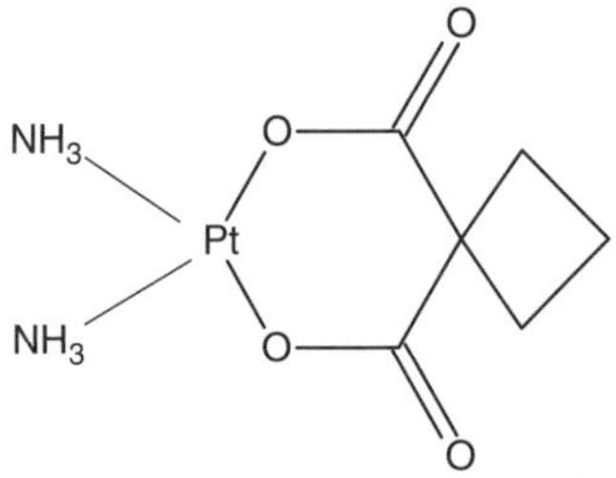

2

[Carboplatin:
diamine cyclobutane dicarboxylate
platinum(II); *cis*-diamine-1,1-cyclobutane-
dicarboxylatoplatinum(II); *cis*-diamine-
(cyclobutane-1,1-dicarboxylato)-
platinum(II)]

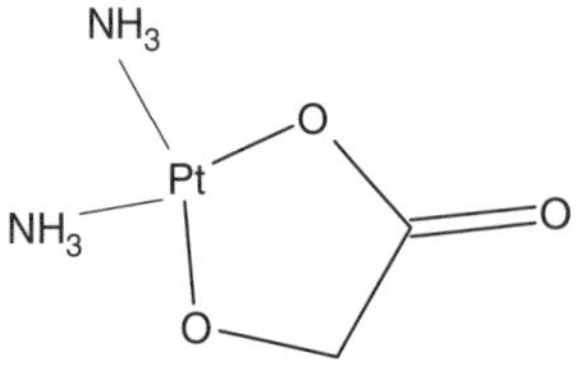

3

[nedaplatin:
cis-diamineglycolatoplatinum(II)]

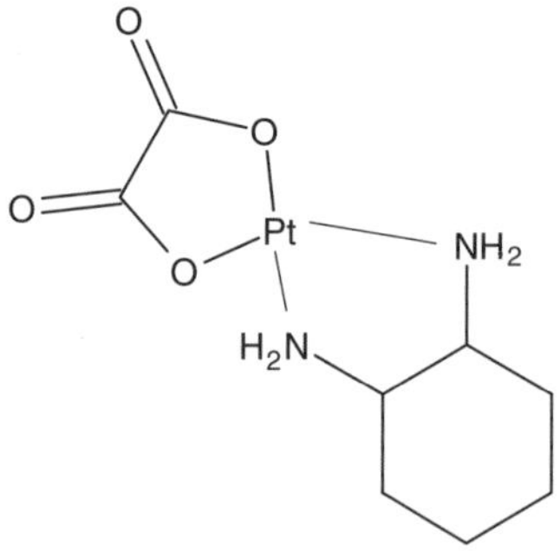

4

[oxaliplatin:
oxalatoplatinum(II)]

Next to cisplatin (**1**), carboplatin (**2**) is the most widely used metal-containing anticancer drug. While it is similar to cisplatin in its cell-killing ability, it shows moderate effectiveness with some malignancies that are less responsive to cisplatin such as non-small-cell lung cancer. It offers a different pharmacokinetic behavior. The presence of the bidentate carboxylate moiety gives decreased rates of reaction with the biological environment. Thus, it shows less nephrotoxicity and is preferred for patients suffering from kidney failure. It also shows a reduced rate of serum protein binding with only 10–20% irreversibly bound to protein. This results in greater bioavailability and larger concentrations of the administered drug to about a fivefold extent. There are a number of good reviews covering carboplatin and related drugs.[47-50]

V. PROPERTIES OF CISPLATIN

In aqueous solution cisplatin is known to undergo spontaneous hydrolysis. (e.g., see Refs. 7, 11, 52, 53). The reaction produces species such as monoaquo platinum and diaqua platinum complexes arising from nucleophilic substitution in water.

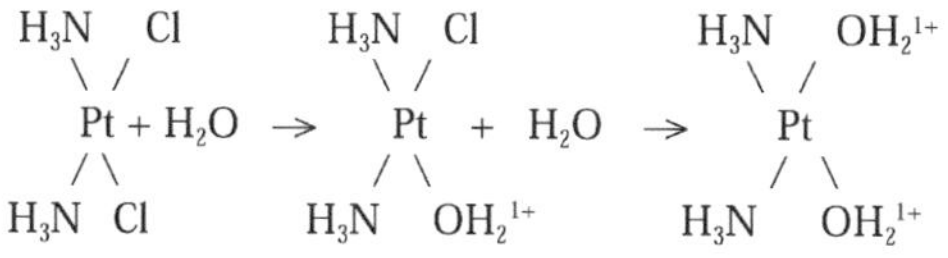

$$
\begin{array}{c}
H_3N \quad Cl \\
\backslash \; / \\
Pt + H_2O \\
/ \; \backslash \\
H_3N \quad Cl
\end{array}
\rightarrow
\begin{array}{c}
H_3N \quad Cl \\
\backslash \; / \\
Pt \\
/ \; \backslash \\
H_3N \quad OH_2^{1+}
\end{array}
+ H_2O
\rightarrow
\begin{array}{c}
H_3N \quad OH_2^{1+} \\
\backslash \; / \\
Pt \\
/ \; \backslash \\
H_3N \quad OH_2^{1+}
\end{array}
\qquad (11)
$$

In addition to the aqua species noted above, other aqueous species can exist, including hydroxy complexes such as **5** and **6**. The actual form of the hydroxyl species is pH-dependent. At a pH of 7.4, 85% of the monohydrated complex will exist in the less reactive dihydroxy form (**6**). Lowering the pH to 6.0 results in the most common form (80%) being the monohydrate species (**5**).

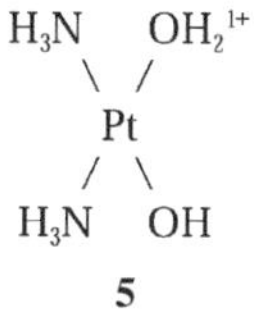

$$
\begin{array}{c}
H_3N \quad OH_2^{1+} \\
\backslash \; / \\
Pt \\
/ \; \backslash \\
H_3N \quad OH
\end{array}
\qquad\qquad\qquad
\begin{array}{c}
H_3N \quad OH \\
\backslash \; / \\
Pt \\
/ \; \backslash \\
H_3N \quad OH
\end{array}
$$

$$\textbf{5} \qquad\qquad\qquad\qquad\qquad \textbf{6}$$

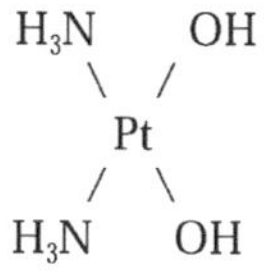

Thus, the possible number of aquated forms derived from *cis*-DDP is great and the proportion dependent on such variations as time, pH, temperature, and concentration of associated reactants (chloride ion and ammonia). Figure 1 contains structures of some of the aquated forms of cisplatin.

The relatively high chloride concentration (about 100 mM) in blood minimizes hydrolysis and the formation of aquated species.[7,11,51,53] Once inside the cell, where the chloride ion concentration is much lower (4 mM), hydrolysis readily occurs, giving a number of aquated species including the diaqua complex. At 37°C the half-life for the completion of the formation of the diaqua complex is 1.7 h with an activation energy of about 20 kcal/mol (80 kJ/mol).[7,11,51,53]

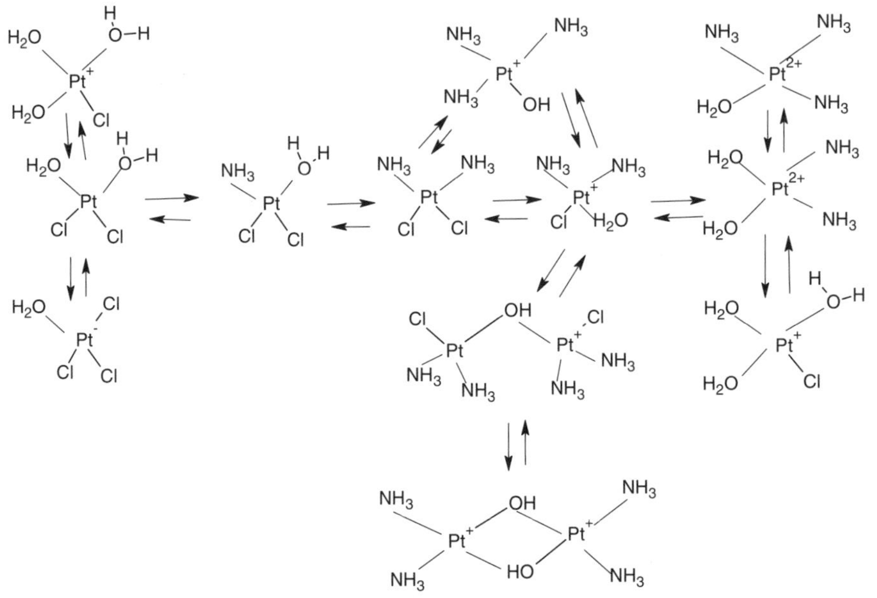

Figure 1 Selected aquated forms of cisplatin.

While the active form within the cell is believed to be the monohydrated structure **(7)**, the "preferred" extracellular species contains two *cis*-oriented leaving groups that are normally chloride ligands. As noted before, due to the high chloride ion concentration in blood, these leaving groups will remain in position causing the molecule to be electrically neutral until it enters the cell.

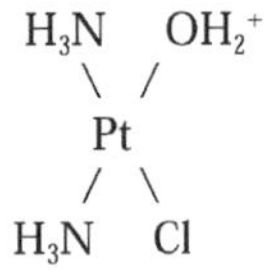

$$H_3N \quad OH_2^+$$
$$\diagdown \diagup$$
$$Pt$$
$$\diagup \diagdown$$
$$H_3N \quad Cl$$

7

As noted above, *cis*-DDP enters cells by diffusion where it is converted to an active form. This is due to the lower intracellular chloride concentration, which promotes ligand exchange of chloride for water and thus formation of the active aquated complex. Thus, the platinum-containing complex should be neutral to enter the cell and labile chloride groups need to be present to form the active species within the cell. The antineoplastic activity of *cis*-DDP appears to be related to its interaction with DNA nucleotides, as a monoaquo species.[13] The monohydrated complex reacts with the DNA nucleotide, forming intra/interstrand crosslinks. Of the four nucleic acid bases, *cis*-DDP has been shown to preferentially associate with guanine. The most common are intrastrand crosslinks between adjacent guanines.[6]

There are several possible crosslinks with DNA. One favored interstrand option occurs between the 6-NH groups of adenines on opposing strands in an A-T-rich region.[11,49–64] This is because these groups are approximately 3.5 Å apart, close to the 3 Å distance between the cis leaving groups on the platinum. The second favored option is crosslinking occurring between the amino groups of guanine and cytosine in opposing strands. This is favored because the platinum is at right angles to the bases that in turn are coplaner with one another. This implies that the bases will have to either "bend down" or "turn edge" to achieve the necessary configuration to bind to the platinum complex. This binding pattern is believed to lead to perturbation of the secondary structure and minor disruption of the double helix. This is sufficient to cause inhibition of DNA replication and transcription with eventual cell death, yet too small to cause a response by damage recognition proteins and consequent excision of the affected segment and repair of the strand.[11,49–69a]

Figure 2 illustrates three of the possible general *cis*-DDP DNA strand cross links.

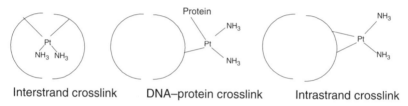

Interstrand crosslink   DNA–protein crosslink   Intrastrand crosslink

Figure 2 General crosslinked forms of DNA from reaction with cisplatin.

Initially, it was believed that atom six of the guanine was the site of importance, being linked with both carcinogenesis and the mechanism of antitumor activity. This belief was amended when tests on guanine-like structures (such as 1,3,9-trimethylxanthine) showed that *cis*-DDP bound to the N^7 and not atom six of guanine. Further studies showed that *cis*-DDP preferentially bonded at N^7 of purines adenine and guanine, and at N^3 of pyrimidines cytosine and uracil.[69a]

While three various types of crosslinks can be formed by cisplatin, the intrastrand crosslinks are more common. Most of these crosslinks can be repaired, but at least one type of interstrand crosslink may not induce response of cellular repair enzymes. Further, while intrastrand crosslinks are important in describing the activity of cisplatin, inducing apoptosis also appears to be a factor in the mechanism of cisplatin's anticancer activity.[69b,c]

While *cis*-DDP is believed to act within the cell, some platinum-containing compounds appear to act on the cell membrane such as the "platinum blues" (e.g., see Refs. 70, 71). Thus, the precise mode and site of activity may vary.

Despite the unquestionable success story of cisplatin, limitations remain, including the powerful toxic side effects.[69a] These toxic side effects include gastrointestinal problems such as acute nausea, vomiting, and diarrhea; occasional liver dysfunction; myelosuppression involving anemia, leukopenia, and thrombocytopenia;

nephrotoxicity, and less frequently cases of immunosuppression, hypomagnesia, hypocalcemia, and cardiotoxicity.[69a]

The most serious side effect is damage to the kidney.[11,72,73] Much of the administered *cis*-DDP is filtered out of the body within a few hours, exposing the kidneys to bursts of high concentrations of platinum. The rapid rate at which the kidneys filter the platinum from the blood is believed to be responsible for the kidney problems. Another problem is the cumulative and irreversible hearing loss experienced first in the 4000–8000 Hz range and then later in the 1000–4000 Hz range. Complete deafness may occur just prior to death.[11,72,73]

VI. STRUCTURE–ACTIVITY RELATIONSHIPS

The topic of structure-activity relationships is complex and not fully defined. While many promising products have been made, overarching structure–activity relationships are uncertain for a number of reasons, not the least of which is that different platinum-containing compounds may inhibit cancer growth in different ways. Thus, the following discussion should be considered as only one brief attempt at describing general structure–activity relationships. First, there should be two available anionic leaving groups such as chloride, bromide, or oxalate. Bidentate chelating groups such as dicarboxylate dianions, except for the especially labile malonato ligand, are often preferable to monodentate ligands because of their superior ability to remain intact in the bloodstream. Complexes with more labile groups such as the nitrate ion hydrolyze too rapidly for in vivo use, while ligands such as the cyanide ion bind the platinum too tightly impairing its activity.[7,11,51,69a] Further, such complexes should have the cis geometry and be neutral with relatively inert amine or nitrogen donor groups. The neutrality of the molecule is believed to allow the platinum-containing drug to more easily traverse the cell membrane. The amines should be primary or secondary amines, allowing for hydrogen bonding to occur.

The relatively high Pt–N bond strength results in tight bonding of the amine, or related, ligands while the leaving groups such as chlorides and carboxylate anions are more weakly bonded and are readily replaced by other nucleophiles. A number of aqua complexes are formed in water and the particular preferred structures are pH-dependent. These aqua structures are in turn susceptible to replacement. When the pH is greater than 6, chloride displacement by the hydroxyl anion is favored, leading to complexes containing the hydroxo group which is a relatively poor leaving group.

Some researchers have used the tetraiodoplatinum dianion because of the more powerful trans effect of the iodine anion compared to the chloride ion.

While most active platinum compounds have these characteristics, there exist active platinum compounds that do not conform to these general requirements. Farrell and coworkers have described three classes of complexes of the general form $PtCl_2LL'$. The structure of one of these where $L = L' =$ pyridine is given as **8**.[7,51,74,75]

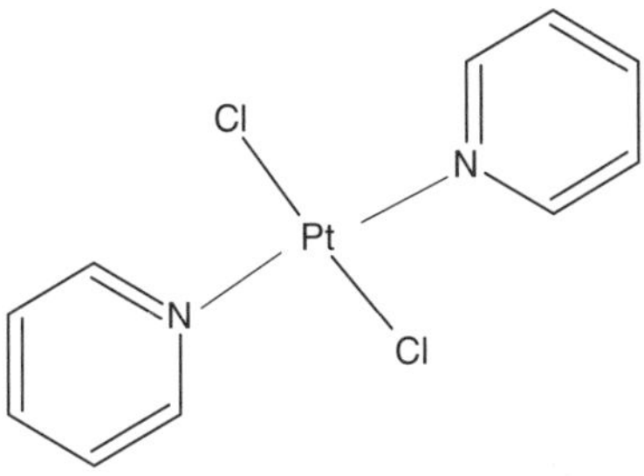

8

These trans complexes displayed good antiproliferative activity. In such complexes a plane can be envisioned through both pyridine rings that might allow for more effective DNA base intercalation than would be achievable with corresponding cis isomers.

Farrell and coworkers also synthesized products of the form *trans*-[PtCl$_2$(het)(R,R$_1$SO)], where het = *N*-heterocycle, R, R$_1$ = alkyl,aryl. When het = quinoline and R,R$_1$ are methyl, the complex shows activity equivalent to that of cisplatin.[76]

In another study, Coluccia et al.[77] found that trans complexes with iminoether ligands, **9**, were more active than the corresponding cis complexes.

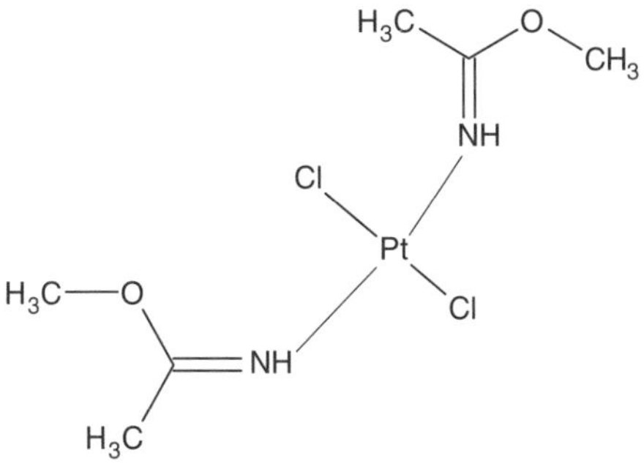

9

Several diplatinum compounds seemingly break the general rules.[7] The platinum atoms in the structures below (**10** and **11**), each have only one ready leaving group. However, they are active against a number of *cis*-DDP-resistant cell lines. These structures are interesting for two reasons: (1) the complex as a whole has two good leaving groups, one chloride on each platinum; and (2) the bridging nature appears to be advantageous and it is similar to what is being attempted in the synthesis of polymers containing platinum in the backbone.

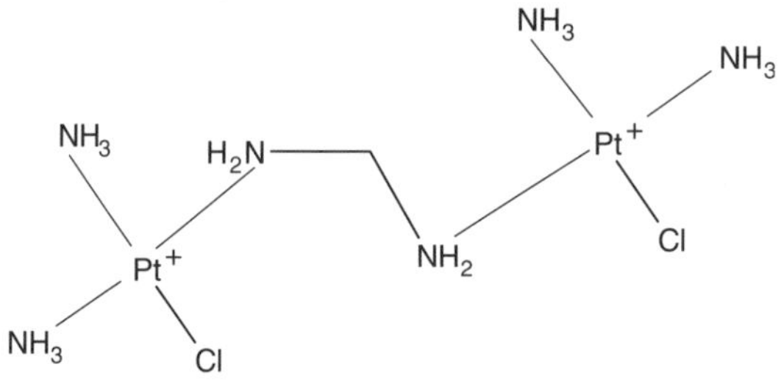

10

A related compound synthesized by Kraker and coworkers[78] has two platinum(II) atoms and four chloride potential leaving groups.[79a]

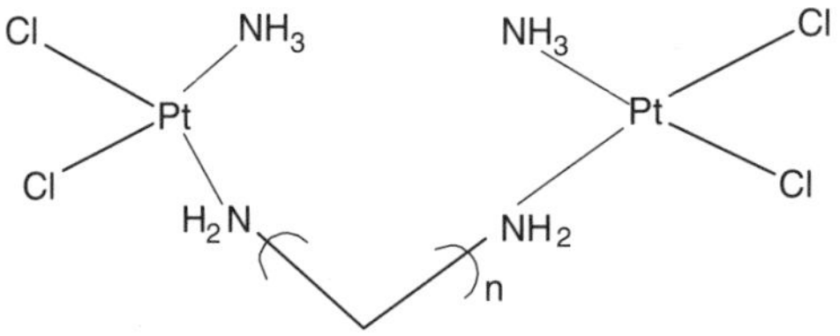

11

This product **(11)** is active against both cultured and xenografted cancer cell lines with the activity dependant on the length of the connective hydrocarbon chain. The greatest activity was found for $n=7$.

While most of the complexes studied employ Pt(II), octahedral complexes with Pt(IV) of the general structure given in **12** have also been found to be potential anticancer drugs. Such complexes must have a low tendency to rearrange and contain axial ligands that are sufficiently susceptible to bioreduction with elimination of the axial ligands and regeneration of the square–planar Pt(II) structure.[7,51]

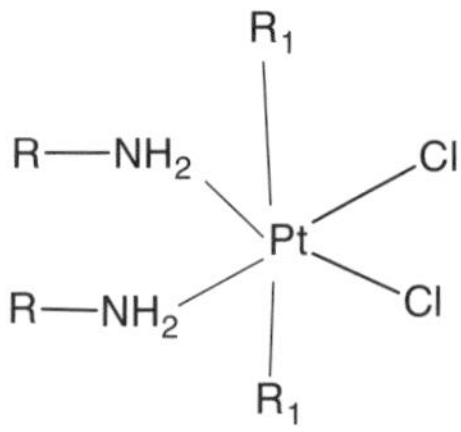

12

Several octahedral compounds have been developed by Kelland and coworkers.[48,79a] One of these compounds is product **13**.

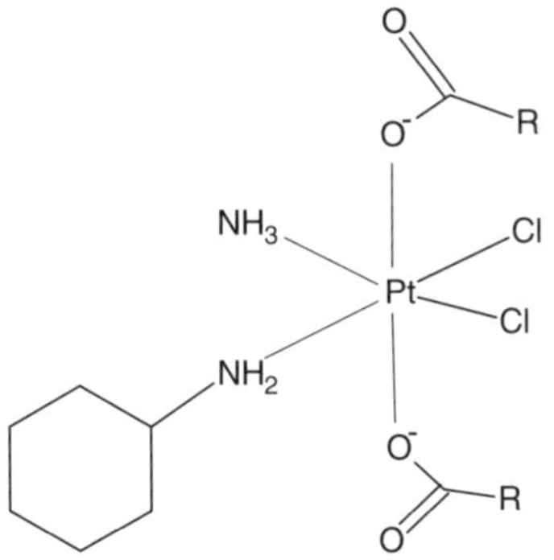

13

Some of these compounds are highly potent in vitro against a panel of human ovarian carcinomas including cisplatin-resistant cell lines. They are also active in vivo when given orally. The most promising structure has an amine and cyclohexylamine grouping. The rationale is that this asymmetric diamine arrangement, along with causing increased water solubility, may also give damaging DNA adduct structures different from those of cisplatin itself. The nature of the carbosylato groups is important with the cytotoxic ability increasing as the R group goes from methyl to butyl for inhibition of cisplatin-resistant cells.

The variety of platinum-containing compounds that exhibit reasonable inhibition of cancer cell lines is large as indicated by the preceding discussion. The structures for potentially useful platinum-containing polymers should also be variable.

VII. POLYMER–DRUG CONJUGATION STRATEGY AND POSSIBLE BENEFITS

Since the discovery by Rosenberg[1] that cisplatin is an effective anticancer drug, large synthesis and evaluation programs have aimed at the creation of cisplatin derivatives that show greater and more widespread activity against cancer but with lowered toxicity. More recent activity has focused on the construction of platinum-containing homing compounds that act specifically at the desired cancer site.

For many decades now, oncologists worldwide have taken advantage of the cytotoxic action of a great variety of drug systems in the fight against cancer. However, despite undisputed successes in cancer chemotherapy, particularly in combination with surgery and other treatment modalities, numerous important pharmacological deficiencies of anticancer drugs have been widely recognized in the medical fraternity. Most drugs lack cell specificity, failing to discriminate between normal and cancer cells. Thus, they tend to cause severe and dose-limiting systemic toxicity. As extraneous agents, they immediately expose themselves as targets for scavenger proteins or as substrates for glomerular filtration and first-pass liver metabolism. As a consequence, serum residence times are commonly short, predominant fractions of administered doses are prematurely excreted (and wasted in the process), and bioavailability (concentration in the target tissue) is generally low. Many drugs are polar, charged, or saltlike. Therefore, they are poor substrates for membrane penetration, intracellular trafficking, and cell entry by the passive diffusion mechanism common to neutral and nonpolar compounds. In addition, drugs possessing poor solubility in aqueous media are sluggishly and incompletely dissipated in the central circulation and become easy targets for the reticuloendothelial system. Finally, and most importantly, acquired drug resistance, which gradually builds up in the target cells after initially successful chemotherapy, is a relatively common phenomenon requiring premature treatment termination. The overall result of these shortcomings is a narrow therapeutic window and grossly limited overall chemotherapeutic effectiveness.

One approach to circumvent these deficiencies is to convert the active agent into some form of prodrug that will encounter minimal interference by scavenger mechanisms. This prodrug will be able to cross intercellular membranes and approach the target site, which in cancer chemotherapy means being able to breach the lysosomal compartment of the cancerous cell.

A. Polymers as Carriers

A brief description is now outlined of the general approach taken by many researchers and described by Neuse and others.

The technology of polymer–drug conjugation stands out as one of the most promising approaches among the strategies being developed and tested out in advanced drug research to attain good drug delivery. Ringsdorf,[79b,c] as early as 1975, articulated the bioreversible binding–conjugating–of a medicinal agent to a macromolecular carrier as a practical biomedical tool that would provide the required prodrug function. In more recent years this concept has been explored and practiced in numerous laboratories, and some excellent reviews covering earlier and more recent work are available.[79b–g]

The typical drug carrier model consists of

- A linear polymer backbone containing solubilizing entities to render it water-soluble
- Functional groups capable of reversible binding to the drug species
- Ideally, some form of "homing device," that is, a functionality showing affinity to the target tissue, providing a degree of drug accumulation in the tumor

The backbone should be

- Nontoxic and nonimmunogenic
- Flexible for solubility enhancement
- Constructed so as to undergo gradual biodegradation, permitting slow fragmentation in the spent state, specifically, after drug release, for efficacious excretion.

The drug-binding functionality should be designed so as to keep the drug essentially carrier-bound while in central circulation, yet allow for its enzymatically or hydrolytically controlled liberation from the carrier in the lysosomal and endosomal space. The conjugate's molecular mass should be in the approximate range of 25,000–80,000; this will retard premature renal excretion, while minimizing toxic effects as occasionally shown by high-molecular-mass polymers.

As a typical prodrug, the water-soluble polymer–drug conjugate serves to introduce the drug component swiftly, and in the perfectly dissolved state, into the aqueous vascular system. In the process, it provides sufficient steric bulkiness to shield the drug temporarily from attack by serum proteins. Extended serum circulation half-life will be the result. Using a pinocytotic cell entry mechanism, the carrier-attached drug, even if polar or charged, will be transported into the intracellular space, thereby overcoming possible influx inhibition, or efflux acceleration, mediated by certain well-defined drug resistance mechanisms. The process will be expedited by the presence of potentially cationic moieties, such as *tert*-amino groups. Cationic sites in a polymer are known to facilitate pinocytosis while, at the same time, increasing affinity for the neoplastic cell, which in many cancers is negatively charged. Finally, the enhanced permeation and retention (EPR) effect[79h] associated with macromolecules, in contradistinction to nonpolymeric compounds, provides preferred distribution of polymers to tumorous tissue. As a consequence, conjugate accumulation in the tumor is favored over that in healthy tissue. The overall result is reduced systemic toxicity and enhanced bioavailability.

Enhanced activity through an increased opportunity for multiple bonding at a given site is also possible for polymers containing multiple binding sites. Further, the presence of the polymer chain should discourage formation of unwanted hydrolysis products while favoring the retention the desired structure.

B. Polymers as Drugs

Another approach involves inclusion of a cisplatin-like moiety into a polymer where the polymer may act as a drug itself, and/or as a controlled release agent similar to that noted above. Here we will focus on the requirements for the use of these polymers as drugs themselves. These requirements are also reviewed elsewhere and share characteristics with the requirements for polymer carriers described earlier.[11,80,81]

It is hoped that inclusion of the cisplatin-like moiety into a polymer will achieve the following:

1. Limit movement of the biologically active drug. Because of their size, polymers are not as apt to easily pass through membranes present in the body. Cisplatin itself is rapidly excreted from the body, causing the kidney and other organs to be exposed to high concentrations of platinum. Polymers with chain lengths of about 100 units and greater typically are unable to move easily through biological membranes. Restricted movement may prevent buildups in the kidneys and other organs, thereby decreasing renal and other organ damage along with associated effects. Further, the platinum from polymers should be released slowly, reducing the exposure of organs to large concentrations of platinum-containing complexes. This should also increase the concentration of platinum in the

beneficial form that is retained in the body, permitting lower effective doses of the drug to be used.

2. Enhance activity through an increased opportunity for multiple bonding interactions at a given site (chemical bonding, hydrogen bonding, hydrophobic interactions, etc.).

3. Increase delivery of the bioactive moiety and decrease toxicity. In aqueous solutions, cisplatin hydrolyzes with a reaction half-life of 9 h at room temperature or 2.4 h at 37°C. Cisplatin hydrolyzes in the body, forming a wide variety of platinum-containing agents, none of which is as active as cisplatin itself and most of which exhibit increased toxicity to the body. Formation of these hydrolysis products increases the amount of platinum complex that must be added to effect desired tumor reduction. Consequently, increases the amount of platinum complexes that must be processed by the body. It is believed that the polymeric nature of the drug will "protect" the active portion through steric constraints, restricting the approach of water to the active site. Also contributing to this protection is the fact that the polymer is not as hydrophillic as the cisplatin itself, as shown by the lack of water solubility of most cisplatin-containing polymers.

4. Bypass the cell's defense system. The cell's defensive response is armed by the invasion of other chemotherapeutic (chemo) drugs. Recent (as of 2003) studies are indicating that introduction of chemo drugs into cells causes the buildup of "housekeeping" proteins that are rather general in their ability to select and remove foreign compounds present in the cell. This may be a principal reason why chemo treatments lead to resistance to drugs, even to drugs not used previously. It is possible that the polymeric nature of the platinum carriers will discourage the housekeeping proteins from removing these polymeric drugs, thus allowing the polymers to function as anticancer drugs under conditions where smaller platinum-containing drugs are not successful.

C. General

Most of these strategies are based in part on the assumption that the platinum-containing polymer will act similarly to cisplatin itself. However, it is likely that at least some of these polymeric drugs act in other ways to inhibit cancer growth. While this complicates the description of overall strategies, it may be an advantage since those drugs would operate in another manner to complement cisplatin as a drug. Thus they may operate in several ways, allowing inhibition of cancer growth to occur via several mechanisms. There is a great need for further work to be done that will allow the particular mechanism of inhibition to be described for these promising drugs.

In the following section we shall first focus on the progress made worldwide in the utilization of the polymer–drug conjugation concept for the enhancement of the platinum drugs' therapeutic efficacy. Carrier-bound drug systems, designed to perform by the rules laid down above, must contain suitable cleavage sites that permit drug release by mechanisms—generally hydrolytic or enzymatic—responsive to the particular biological environment. It stands to reason that these release mechanisms depend crucially on the structural peculiarities of the mode of drug incorporation into the conjugate.

The square–planar cisplatin-type structural complex *cis*(R−NH$_2$)$_2$PtL$_2$ (R=H, alkyl or aryl; L=leaving group, such as halo or carboxylato) can be transformed into a polymer constituent in a variety of ways. In the simplest and most straightforward case, it may function as a mainchain component, for example, through metal interconnection by a diamine moiety. This arrangement, while not in conformance with Ringsdorf's polymer-drug conjugation model, will be discussed first (Section VIII) because it has been investigated in one of the earliest projects in the platinum polymer field. Metal interconnection offers a great number of possible synthetic approaches and structural variations. Further, it illustrates the use of polymers as drugs themselves.

In a related mode, the metal, incorporated via Pt−N bonds as before, yet no longer functions as a mainchain constituent. Instead, it is coordinated by one or two nucleophilic nitrogen donors, generally amino groups, preintroduced into presynthesized carrier polymers. The Pt−N anchoring mode will be the topic of Section IX.

In still another polymer-binding mode, the metal is coordinated by carrier-attached ligands of the leaving group type, notably carboxylato or hydroxylato ligands. This important case of platinum coordination will be discussed in Section X.

VIII. MAINCHAIN-INCORPORATED *cis*-DIAMINE-COORDINATED PLATINUM

A. Simple Amine Derivatives

A major effort at utilizing the platinum-containing polymer itself as a drug is the mainchain incorporation of platinum based on the de novo synthesis of the polymer structure from specified monomers, as no presynthesis of carrier polymer is required. The preparative approach is exemplified by the polymerization of tetrachloroplatinate(II) dianion with diamines and other nitrogen-containing compounds.

The initial polymers containing Pt−N in the backbone were reported in 1981. The reaction was based on the synthesis of *cis*-DDP itself, except the nitrogen-containing reactants were diamines. Much of this work was reviewed in 1985.[11]

Initially both aromatic and aliphatic diamines were used as well as amino pyrimidines, including a thio amino pyrimidine.[11,82–95] Amino pyrimidine structures were thought to be readily recognizable to the cells. The rational for the use of a thio pyrimidine was related to the general need by the body for sulfur-containing compounds. Both ploys were intended to encourage the cells to interact with the platinum-containing polymer. Along with the dichloro derivatives, diiodoplatinum(II) derivatives were also synthesized.

Synthesis was straightforward, accomplished by mixing together aqueous solutions containing the platinum salt and the diamine. Reaction was rapid; the polymer began to precipitate from the reaction mixture in minutes.

The products were polymeric with molecular weights in the range of 10^6 daltons. The products were soluble at low concentrations in DMSO. The products showed 100% inhibition of HeLa and L929 cell lines at a concentration level of 50 μg/mL.

The viral replication of poliovirus type 1 in HeLa cells was also studied, with the product from tetrachloroplatinate(II) and 1,6-hexanediamine (14) able to effectively inhibit viral growth at a concentration level of 10 µg/mL.[82] (The viral activity is discussed further in Section VIII.F.) The compounds were relatively nontoxic, with mice able to tolerate 400 µg/mouse (highest level tested) without noticeable ill effects.[82]

14

The polymeric derivatives are believed to exhibit lowered toxicities in comparison to cisplatin itself. An often employed upper dosage of cisplatin (single dose) is about 4×10^{-4} g/kg for humans, which is increased to 4×10^{-3} g/kg when flushing is employed. For rats the LD_{50} (ipr-rat) is 1.2×10^{-3} g/kg. Mice have been injected on an alternate-day schedule for one month with DMSO-H_2O (10%–90% by volume) solutions containing 2×10^{-2} g/kg of the polymer derived from tetrachloroplatinate (II) and 1,6-hexanediamine without apparent harm.[87]

The cytopathic effect of platinum(II) polyamines on normal and transformed 3T3 cells was studied.[84,86] Some of the compounds were more toxic to normal cells, while others were more effective at destroying the transformed cells. A cell differential ratio, which was simply the ratio of the percentage of cell death for the transformed cells compared to the normal cells, was established to rate the relative effect of cell inhibition. A value greater than one indicated that the polymer was more effective against the transformed cells. Examples of polymers with low cell differential ratios were products derived from tetraiodoplatinate, PtI_4^{2-}, and 1,8-diamino-*p*-methane and the product from tetraiodoplatinate and pyrimethamine. The product with the largest cell differential ratio, a ratio of 14, was derived from tetrachloroplatinate and 2-chloro-*p*-phenylenediamine.

The second highest cell differential ratio, a cell differential ratio of 4, was observed from the product of tetrachloroplatinate and methotrexate. Methotrexate is a cancer drug itself. The product illustrates the advantage of connecting platinum-containing units onto other cancer drugs. This polymer was tested, with a few others, by Bristol Laboratory for the ability to inhibit L1210 leukemia in mice, utilizing a value called the "%T/C," which is defined as the median survival time for treated mice divided by the median survival time for the control times one hundred. The methotrexate polymer, **15**, showed a %T/C of 164 at a dosage of 16 mg/kg.[11,84]

There was a correlation between the %T/C values and the cell differential values. Those that showed cell differential values greater than one also showed %T/C values greater than one, and those showing low cell differential values also showed low % T/C values.

15

Two of the platinum polyamines were tested more extensively against normal and transformed cells.[88,89] Polymers in which the diamine component was methotrexate or 2-chloro-*p*-phenylenediamine were used to treat a variety of mouse Balb/3T3 cells, including normal nontumorigenic cells, spontaneously transformed tumorigenic cells, and virus-transformed tumorigenic cells. When treated with the polymer containing 2-chloro-*p*-phenylenediamine, all cell types were actively killed at a concentration of 10 µg/mL. Cell growth was inhibited 50% at a concentration of 5.0 µg/mL. However, normal quiescent Balb/3T3 cells were unaffected by polymer concentrations as high as 20 µg/mL, indicating that cells must be actively growing in order for the polymer to exert its effects. This would be consistent with a mechanism of polymer action involving the inhibition of DNA replication, similar to that of cisplatin. The polymer containing methotrexate gave similar results, except that this polymer was far more active, killing the various cell types at a concentration of 0.5 µg/mL and inhibiting cell growth 50% at a concentration of 0.06 µg/mL. Again, normal quiescent cells were resistant to the polymer. The stronger activity of this polymer demonstrates the advantage of combining two different anticancer agents in one compound.

To investigate the structure–activity relationship of the platinum polyamines, about two dozen different polymers were synthesized, each with a different diamine component. The diamines varied from relatively simple ones, such as 2-nitrophenyl-diamine and 2,5-diaminopyridine, to rather complex structures including tilorones and acridines. The biological activity of these polymers was tested using Balb/3T3 cells, and a wide range of activity was found. Some polymers were essentially inactive, showing no effect on cell growth at concentrations up to 50 mg/mL. The diamine components of these inactive polymers included amino acids and pyridine derivatives. Other polymers were extremely active, significantly inhibiting cell growth at concentrations of 0.1–1.0 µg/mL. The diamine components of these very active polymers included proflavin, euchrysine, acridine yellow, and tilorone, as well as methotrexate. In addition to affecting normal Balb/3T3 cells, these polymers also inhibited the growth of various human ovarian cancer cell lines.

The more active polymers were comparable to cisplatin in their biological effects, killing or inhibiting the growth of a variety of actively growing cells at similar concentrations, but having little effect on quiescent cells. In one respect, however, the polymers differed significantly from cisplatin.[90] The activity of a cisplatin solution depends on the age of the solution. While fresh solutions of cisplatin in DMSO or water are very active, while solutions stored one week, even at 4°C, are 100-fold less active. This is due to reaction of the cisplatin with the solvent, forming substituted charged species that cannot easily enter the cell. In contrast, the age of a polymer solution does not affect its biological activity. Polymer solutions in DMSO stored for one week at room temperature showed the same effects on cells at the same concentrations as did freshly prepared solutions.

An attempt was made to correlate the activity of the platinum polyamines with various factors in order to determine the best characteristics of a potentially useful anticancer agent.[91] The activity of the various polymers was compared with the biological activities of the diamine components themselves. No direct relationship could be seen. In some cases, the diamine component by itself had less biological activity than did the corresponding polymer containing that diamine. In other cases, the diamine component was more active than the corresponding polymer. The structure of the diamine also had no obvious relationship to polymer activity. Some of the active polymers contained relatively small diamine components, while other active polymers had relatively large diamine components. There was also no direct correlation with the hydrophobic character of the diamine component. Although a hydrophobic diamine component might allow a polymer to more easily cross the cell membrane and so be more effective, no such relationship was evident.

Another factor that was explored was polymer size. The molecular weights (degrees of polymerization) of the polymers were determined by light-scattering photometry and size-exclusion chromatography. Polymer molecular weight varied from 7000 daltons (DP=11) for a methotrexate-containing polymer to 600,000 (DP=1300) for a polymer containing 4,4'-diaminodiphenylsulfone as the diamine component. Most polymers had molecular weights in the range of 25,000–125,000 daltons with DP values of 30–200. However, there was no correlation between the size of the polymer and its biological activity. Although a smaller polymer might pass through the cell membrane more easily and so be more effective, some larger polymers were very active as well as active small ones.[92]

Some interesting aspects of polymer size were found. First, the size of the polymer varied with time and reaction temperature, generally increasing in size with longer reaction times and higher temperatures. However, variations in size had no effect on the biological activity of a particular polymer. Second, some polymer products were found to contain two different size classes of chains. The smaller size class was present with short reaction times. As the reaction time increased, the larger chains appeared in increasing proportion to the smaller class. The two different size classes of chains suggest two modes of polymer synthesis. It is believed that the first mode involves synthesis in solution where the polymer molecule grows to a certain size depending on such factors as solubility. Once precipitated, a second mode of synthesis could occur in the solid phase as precipitated chains react together to form longer chains.[93]

As noted elsewhere, some experimenters will utilize tetraiodo platinum in place of tetrachloroplatinate because reaction is much faster. The iodo product is then converted to the chloro product through addition of silver nitrate. This sequence is shown in Scheme 1.

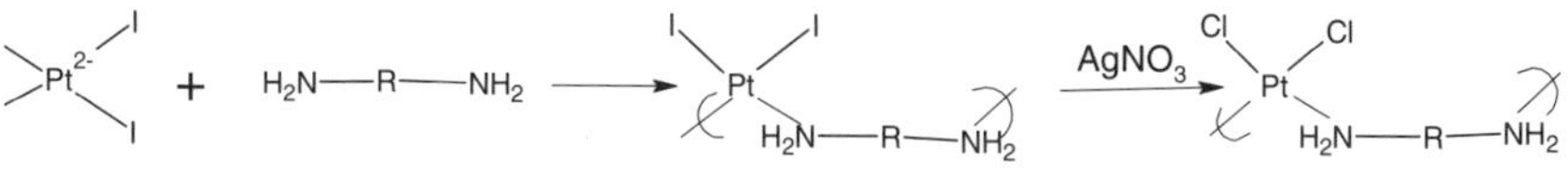

Scheme 1 Formation of the dichloro derivatives via the iodio derivatives.

B. Amino Acid Derivatives

A number of platinum polyamines have been synthesized from reaction with diamino-containing mono, di, tri, and tetrapeptides.[95,96] Diaminomonopeptides (single amino acids containing two free amine groups) were generally not chosen except where the opportunity for cyclization is not favored. Sample amino acid–containing reactants included glycyl-L-histidine, folic acid, carnosine, tryptophan, and D-biotin. Degrees of polymerization ranged from 16 for carnosine to 80,000 for tryptophan. As will be described below, many of these products also contained Pt–O linkages. The Pt–O linkages are formed subsequent to formation of the initial diamine linkages on the platinum for some of these materials. Elemental analysis indicated that these products probably are formed by displacement of one of the chlorines. Thus, the repeat unit for these structures contains a platinum atom surrounded by two nitrogens, one chlorine, and one oxygen.

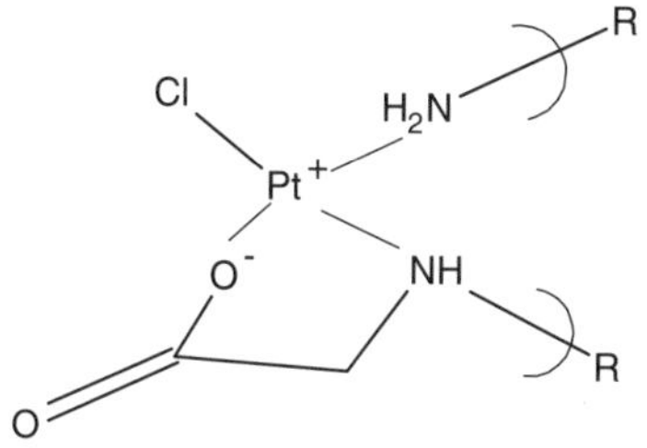

16

One in-depth study involved the formation of *cis*-DDP derivatives based on the reaction with histidine (**17**).[98] This study will be described in some detail because it is important to understand the nature of the reaction and products when working with amino acids and other related amine-containing compounds.

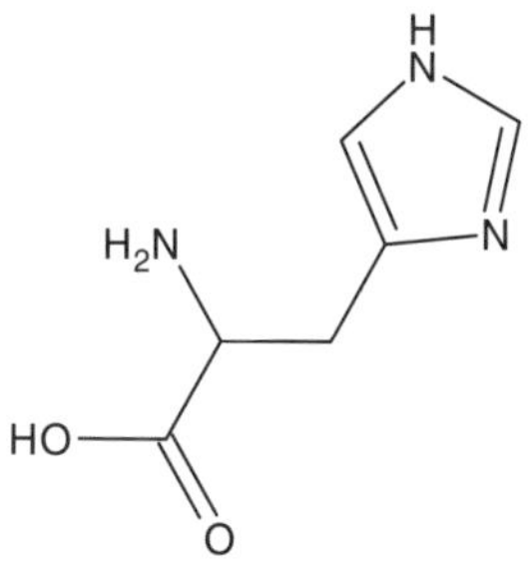

17
(histidine)

The structure of histidine varies with pH as a result of protonation at nitrogen. The general availability of the three nitrogens to bind with platinum depends on this protonation equilibrium, which depends on pH. Further, α-amino acid complexes with heavy metals are well known and these complexes give a 5-membered ring rather than chain extension. Such chelation has a greater likelihood of forming at higher pHs where the amine retains a free pair of electrons available for binding and the carboxylic acid is in the salt form.

When only the reactants are added to water, the reaction pH is about 2–3 and the reaction is complete after 3–4 days. Here the carboxyl group is largely protonated. When an equivalent of base is added to the histidine in this reaction, the pH started at about 7.2 and decreased again to about 4. Here the carboxyl group is largely deprotonated (the pK_a for COOH is about 2.0 for the free amino acid).

The low pH of these reaction solutions indicates that protons are being produced in the system. There are several possible sources. First, displacement of a proton from one of the protonated functional groups in the amino acid by the metal ion is necessary for coordination at the metal site. The reactions of K_2PtCl_4 with amino acids has been extensively studied by Volshtein and others.[99,100] According to Volshtein, reaction of the simple amino acid in the zwitterion form, AH, proceeds according to the following equations:

$$K_2PtCl_4 + AH \longrightarrow K[PtACl_2] + KCl + HCl \qquad (12)$$

$$K[PtACl_2] + AH \longrightarrow [PtA_2] + HCl + KCl \qquad (13)$$

In the zwitterionic form, basic histidine is only protonated at the alpha amino group. Protons generated from this process then result from deprotonation of the third functional group, leaving this site available for coordination with the platinum. In this case, decreases in the pH are an indication of the extent of the reaction.

Anionic histidine is fully deprotonated. Thus, the protons must come from another source. An additional source is through the hydrolysis of K_2PtCl_4. While rarely isolated, the formation of hydroxy complexes is known to be responsible for the acidities of aqua complexes produced by the hydrolysis of K_2PtCl_4 as follows:

$$[PtCl_4]^{2-} + H_2O \longrightarrow [PtCl_3(H_2O)]^{1-} + Cl^{1-} \longrightarrow$$
$$[PtCl_3(OH)]^{2-} + H^{1+} \qquad (14)$$

The first reaction begins after several hours, but replacement of a second chloride ion takes much longer.[101] The hydroxy complexes are seldom isolated because hydroxide ions, being hard bases, do not form stable complexes with Pt(II) ions.[102] Thus, water can displace chlorine ligands from the metal ion, but nitrogen donors, or oxygen donors when part of a chelate, readily displace the aqua and hydroxy ligands. Finally, protons may be produced from hydrolysis of the products formed in the reaction. The production of protons in these reactions was believed to be caused by a combination of proton displacement from the amino acids and hydrolysis of K_2PtCl_4, as a result of the complexation reactions.

Infrared spectra of solids were consistent with a variety of products being formed. These include coordination between the Pt and both the amino and imidazole nitrogens. Further, coordination through the carboxylate groups also occurs. Thus, individual polymer chains will contain a mixture of platinum-coordinating units.

Only a few of the amino acid compounds have been tested. The product from tryptophan showed no activity to 50 µg/mL against Balb/3T3 and transformed Balb/3T3 cells and histidine compounds showed no activity to 25 µg/mL.[91] Clearly, more study is needed to evaluate this potentially important group of compounds.

C. Other Nitrogen–Platinum Products

Polymers can be formed using a number of nitrogen-containing compounds other than diamines. Following is a brief description of some of these other products. All of these have structures analogous to those of the polyamines where the nitrogen units are connected in a cis geometry to the platinum. Further synthesis was also similar, occurring in aqueous solution with the products precipitating from the solution.

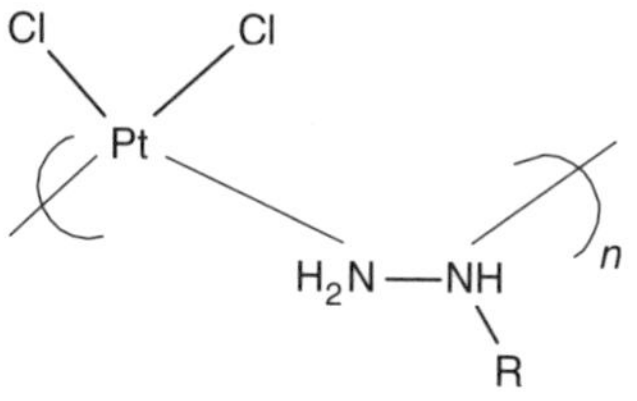

17a

Reaction with hydrazines gives products of structure **17a**.[103]

The products were polymeric with degrees of polymerization in the range of 200–3000. In general, the reaction rate decreases as the electron density on the nitrogen decreases, as expected. The reaction was studied as a function of various substituted phenylhydrazene derivatives and the rate follows the Hammett *para*-sigma value such that more positive the *para*-sigma groups (corresponding to a decrease in the electron density) lower the reaction rate. The trend agrees with other linear-free energy relationships so no importance should be placed on the type of linear-free energy relationship utilized.

The reaction appears to be second order during the initial growth state with specific rate constants similar to those found for reaction with simple monoamines. After the initial relatively rapid growth period, product formation slows, but proceeds through several days reaction time. Reaction yields after 24 h varied from about 14% for phenylhydrazine to 70% for 4-nitrophenylhydrazine. Two tumor lines were studied. Both the L929 and HeLa were inhibited at a concentration of 12 µg/mL.

A number of urea, thiourea, amide, and thioamide products were synthesized.[104,105] The structure for the products derived from thiourea is given in **18,** and the structure for products from oxamide is given in **19**.

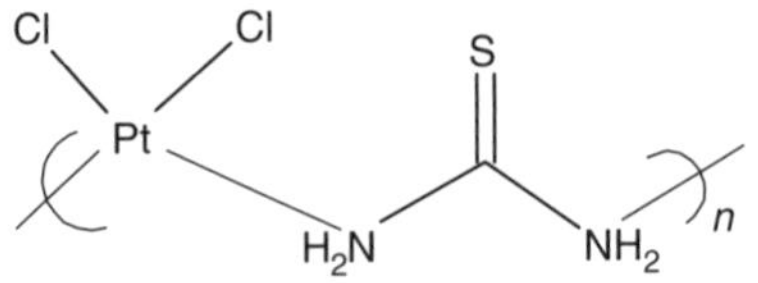

18

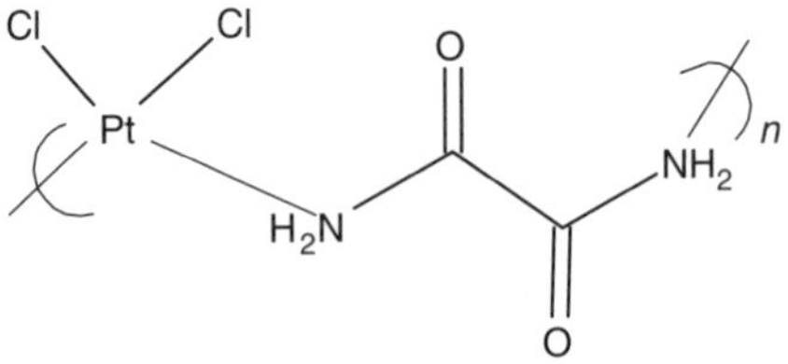

19

While the products could take on an octahedral structure, infrared evidence was consistent with a square–planar arrangement about the platinum atom.

The chain lengths ranged from about 40 to 750. Interestingly, the chain length increased as the electron density on the amine increased. Thus, the urea product had a degree of polymerization of 37 while the corresponding thiourea product had a degree of polymerization of 750.

While the products exhibited some tumor activity against L929 and HeLa cells, the activity was less than that of products derived from diamines. About 50% inhibition of both cell lines occurred at a polymer concentration of 10 μg/mL.

D. Solution Stability

Stability of biological activity was not the same as the stability of the structure. When dissolved in DMSO/water, most of the polymers underwent random chain scission during the first 24–48 h, eventually stabilizing with a size about one-half to one-tenth its original size. This size change did not noticeably affect the biological activities of the polymers.

Molecular weight was studied as a function of time for the product of tetra-chloroplatinate and histidine.[98] After dissolution, which occurred fairly rapidly, the initial molecular weight for one sample was 1.25×10^5 daltons. At day 2 this fell to 6×10^4 and by day 10 to about 4×10^4. Thus, molecular weights need to be watched because chain size is time-dependent with a short half-life.

The degradation of the polymer from methotrexate deserves special attention because of its good activity and high cell differentiation value. As already noted, ready degradation occurs such that after one day polymer chain lengths are lowered by a factor of 10. The mechanism of chain degradation may also be informative: (1), it will give an indication of the mobility, location, and half-life of the polymer in the body and (2) determining the type of degradation should allow better structure tailoring to achieve desired release rates of monomeric and/or oligomeric degradation products. The two main degradation modes are random scission and chain degradation. In random scission, chain rupture occurs at random points along the chain, leaving fragments that are large in comparison to monomeric units. In chain degradation, the chain unzips from one or both ends, creating monomeric units as chain degradation progresses. The results with the methotrexate–tetrachloroplatinate polymer are consistent with degradation occurring through random chain scission.[106,107]

Ready degradation is both positive and negative. On the positive side ready degradation results in availability of the drug. On the negative side, excessively rapid degradation defeats some of the potential advantages of polymeric drugs.

E. Thermal Stability

The thermal stability of a number of the platinum-containing polymers has been studied (e.g., see Ref. 108). Following is a summary of the findings. Some of the platinum polyamines have good thermal stabilities. In air and nitrogen most of the products (often better than 70% by weight) remain to 1200°C. This makes simple elemental analysis by automated thermal methods difficult. Degradation can occur through three different routes:

1. Degradation can occur through nonoxidative routes where the differential scanning calorimetry (DSC) and thermal gravimetric analysis (TGA) thermograms will be identical in air and under an inert atmosphere such as nitrogen.
2. Degradation can occur at the organo portion away from the platinum metal atom. Here the DSC thermogram will show mildly endothermic/exothermic behavior and be different in air and nitrogen.
3. Degradation in air can occur through oxidation at the platinum metal. Here the DSC thermogram will reflect highly exothermic behavior and again, be different in air and nitrogen.

Compounds that undergo degradation via either of the first two routes, such as glycylhistidine, give poor C,H,N elemental analysis results and generally high residue weights to 900°C. Polymers from other diamines such as methotrexate degrade in air through highly exothermic reactions as indicated by DSC. These materials give reasonably good C,H,N elemental analysis results. It is possible that the localized energy created by oxidation at the platinum atom encourages release of the organic portion, giving ready degradation of the material. In other situations, the energy that is given off because of oxidation occurring at the metal site results in the formation of crosslinked products, which are, in turn, more thermally stable. Thus, there exists a connection between the mode of degradation determined by DSC and the elemental analysis results.

F. Antiviral Activity

Some cancers are believed to have a viral relationship. As such it is informative to look at the viral response to some of the platinum polymers.

Tetramisole is an antihelmentic that acts on the cyclic nucleotide phosphodi-esterases. It actually consists as a combination of optical isomers; the most active one is levamisole. Levamisole was the first synthetic chemical that exhibited immunomodulatory properties. It appears to restore normal macrophase and T-lymphocyte functions.

Cisplatin polymer analogues, made through reaction of tetrachloroplatinate with tetramisole, were tested for their ability to inhibit EMC-D viruses that are responsible for the onset of juvenile diabetes symptoms in ICR Swiss male mice.[109] Briefly, the mice were treated with 1, 5, and 10 mg/kg. Doses of 1 and 5 mg/kg decrease in the severity and incidence of virus-induced diabetes in comparison to untreated mice. In another series of tests, doses of 1 and 10 mg/kg were administered one day prior to injection of the virus, but here there was an increase in the severity and incidence of virus-induced diabetes. Other studies were undertaken, showing that the polymer showed different activity profiles than the tetramisole (20) itself.

20
(tetramisole)

In a related study, the methotrexate polymer (15) was similarly tested. In the initial study female mice were treated. Generally, only male ICR Swiss mice develop diabetes-like symptoms. Female mice must first be treated with testosterone before they can develop diabetes-like symptoms. The mice were divided into three groups, all of which received injections of the polymer [0.5 mL intraperitoneally (IP) of polymer solution containing 6.4 mg/kg]. A week later groups I and II were treated with testosterone. On day 8, group 1 was again given a second 0.5 mL IP inoculation of the polymer solution. On day 9, all groups received 1×10^4 pfu (plaque-forming units) of EMC-D virus. On day 17, all mice were given a one-hour glucose tolerance test. The glucose level for groups I and III were similar and significantly below the level for the diabetic mice in group II. This is consistent with the polymer effectively blocking the diabetogenic effects of the virus. Further, other results from this study were consistent with this strain of female mice being susceptible to developing diabetes-like symptoms even without the testosterone treatment.

A related study was carried out using male mice. Here, again, the polymer (15) showed a greater positive effect on the control of diabetes than either of the reactants themselves. The glucose levels were near those of noninfected mice for the polymer-treated mice. Again, the incorporation of both the platinum and methotrexate into a polymer is the effective agent and not either of the drugs themselves. These two experiments are related to generation of a vaccine that can be employed to prevent onset of β-cell damage by RNA viruses.

The third experiment focused on treatment subsequent to viral infection.[110,111] The polymer was 100 % effective in viral control with delivery of the polymer (15) one day after the mice were infected. In summary, the methotrexate polymer (15) is an effective antiviral agent against at least the EMC RNA virus.

A number of platinum polyamines were tested for antiviral activity in tumor cells.[112] For instance, the polymer from tetrachloroplatinate and 2,6-diamino-3-nitroso-pyridine, **20a**, which exhibited a cell differential ratio of 3.4, was tested at a concentration of 2.2 µg/mL on L929 cells infected with Encephalomyocarditis, EMC, virus, strain MM. A virus reduction of about 25% was seen. This is considered to be a moderate antivirial response.

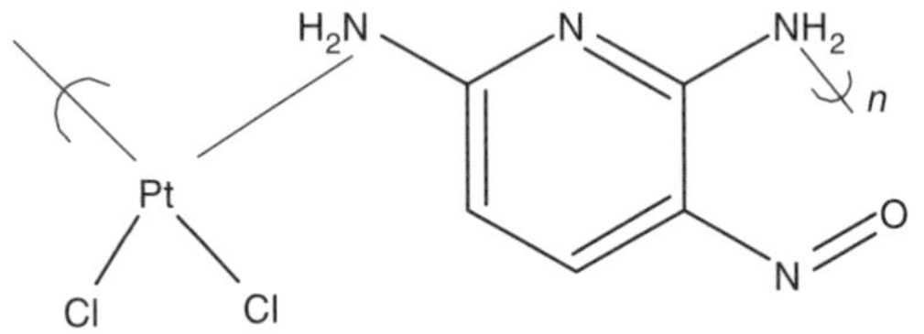

20a

In general, agents capable of inhibiting one RNA virus will inhibit other RNA viruses, but each DNA virus must be evaluated separately. The platinum polyamines were studied against RNA viruses. The behavior toward RNA viruses was varied with some showing little activity but the majority showing inhibition of viral replication at polymer concentrations below which tumoral inhibition is found (<1 µg/mL).

The effect of platinum polyamines on the transformation of 3T3 cells by SV40 virus was also studied.[112] In summary, the polymers showed no effect on the transformation process.

IX. PLATINUM CARRIER-BOUND COMPLEXES VIA NITROGEN DONOR LIGANDS

A. Pt-Polyphosphazenes

One of the earliest investigations aimed at polymer-anchored platinum complexes in medicine falls under this heading. Arguing that polymer-binding of cisplatin might retard or prevent platinum excretion through the kidney filtration system, Allcock and coworkers treated poly[bis(methylamino)phosphazene] with cisplatin in chloroform solution in the presence of a crown ether and so prepared a water-soluble, antitumor-active conjugate. In this conjugate, the $PtCl_2$ moiety was assumed to be polymer-bound via skeletal (mainchain-contained) N atoms based on model reactions. Following is the reported structure of the product of reaction of tetrachloroplatinate(II) and octamethyl-cyclotetraphosphazene (**21**):

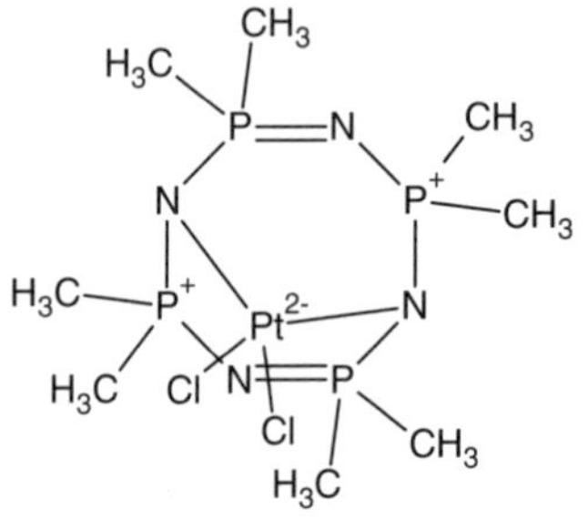

21

An alternative structure is given in **22** where the ligand adds rather than sub-stitutes; that is, a coordination rather than condensation reaction takes place. Ring resonance is retained in **22**.

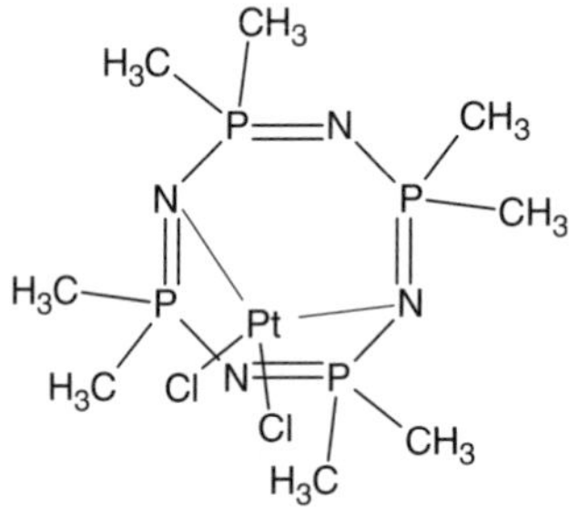

22

The polymeric derivatives of substituted polyphosphazenes have been identi-fied as having structure **23,** where the platinum is linked with the backbone nitrogen rather than at the nitrogen groups attached to the phosphorus. The true structure probably involves a combination of several different structural units.

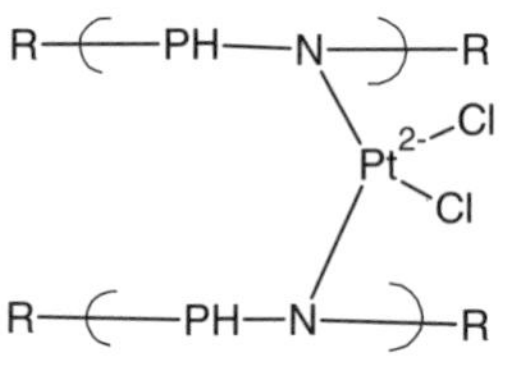

23

Again, another structural unit that could exist for this product is **24**.

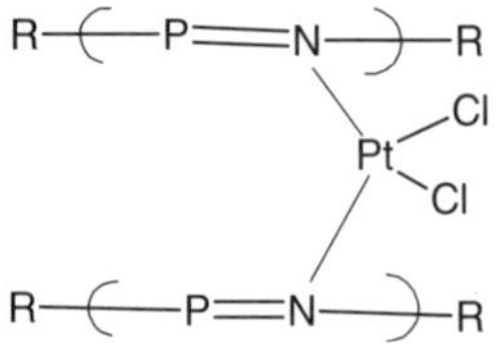

R—(—P=N—)—R
                    \
                     .Cl
                  Pt
                 /   `Cl
R—(—P=N—)—R

24

The products were tested by the mouse P388 lymphocytic leukemia survival test and by the Ehrlich ascites tumor regression test.[113] The polyphosphazene–platinum product showed an inhibition of 86% in the ascites test and a 5/7 survival after the eighth day for the P388 mouse test.

Other individuals have also employed polyphosphazenes in the synthesis of platinum(II) carriers. For instance Song and Shon[114] synthesized diamineplatinum(II) polyphosphazene–carbohydrate conjugates from reaction of polydichlorophosphazene with 5-methoxycarbonyl-1-pentanol forming poly[(5-carboxy-1-pentoxy)(methoxy) phosphazene] linked via L-glutamate spacers with cyclohexanediamine Pt(II), a liver-targeting carbohydrate moiety [*N*–(4-aminobutyl)*O*-alpha-D-galacotopyranosyl-(1->4)-D-gluconamide], and a solubilizing group 2-(2-aminoethoxy)ethanol.

B. Slowly Biofissionable Pt–N Complexes Anchored through Primary and Secondary Amines

Carrier-bound drug systems designed to perform as described in Section VII. A should contain suitable cleavage sites permitting drug release. These release mechanisms depend crucially on the structural peculiarities of the type of drug incorporated into the conjugate.

The square–planar cisplatin-type structural complexes can be transformed into a polymer constituent in a variety of ways. In the simplest and most straightforward case, it can function as a main chain component, for example, through metal inter-connection by a diamine moiety as described in Section VIII. In a related mode, the metal, incorporated through Pt–N bonds that are not part of the mainchain constituent. Platinum is coordinated by one or two nucleophilic nitrogen donors, generally amino groups, preintroduced into presynthesized carrier polymers. This type of Pt–N anchoring mode is the topic of the present section.

Whereas in Allcock's work tertiary, phosphazene-type nitrogen atoms serve as the donor ligands, the more common Pt–N anchoring approaches utilize primary or secondary amine functionalities as the nitrogen donor ligands. Structures falling into this category are preferentially composed of a carrier component featuring 1,2-diamine, 1,3-diamine, and 1,4-diamine segments incorporated either as sidechain terminals or as mainchain constituents. Scheme 2 illustrates these anchoring arrangements.

In the schemes below cleavable bonds, typically of the amide or ester type, are indicated by the symbol //.

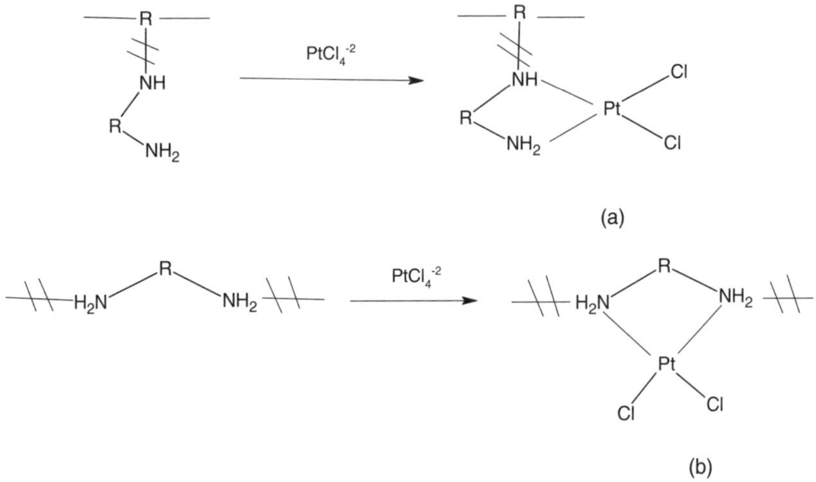

(a)

(b)

Scheme 2 Generalized structures for platinum-carrier polymers derived from sidechain (a) and backbone-containing (b) amine chelating groups.

Also included here are structures comprising carriers bearing monoamino side groups as described in Scheme 3.

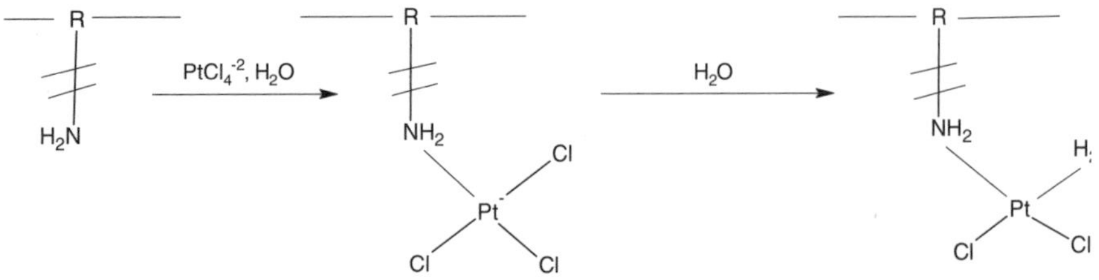

Scheme 3 Generalized structures for platinum-containing carriers employing monoamino side groups.

Depending on the exact structure, the segments can be tightly held and thus less susceptible to biofission or more loosely held with the platinum-containing moiety generally held further from the polymer backbone. Polymers where the platinum-containing moiety is more tightly held will be covered in this section while those that are more loosely held, and thus generally more biofissionable, will be covered in the following section.

Because many of the potential natural carriers contain nitrogen groups that can chelate with the platinum-containing compound, a number of studies were carried out evaluating nitrogen-containing polymers. Further, these products (without bound Pt) were themselves also evaluated as potential anticancer drug materials.

Polyethyleneimine (PEI) has nitrogen groups within its backbone separated by two methylene units. Formation of internal 5 and 6-membered rings is favored by platinum and this arrangement allows the formation of 5-membered rings as shown in **25**.[115-117]

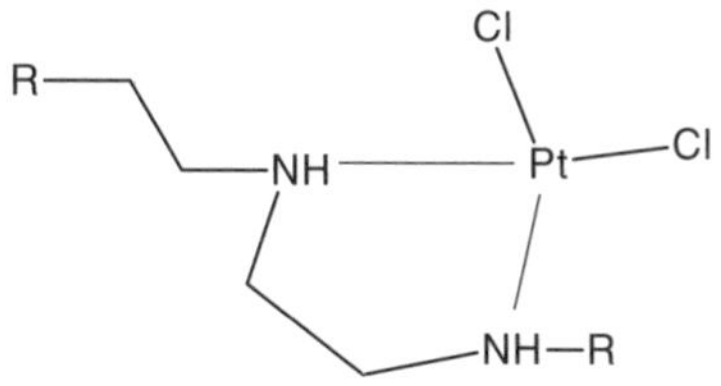

25

Other units are possible including those formed through crosslinking with another chain (**26**) or through reaction between two nitrogens contained within the same chain but somewhat further removed from one another.

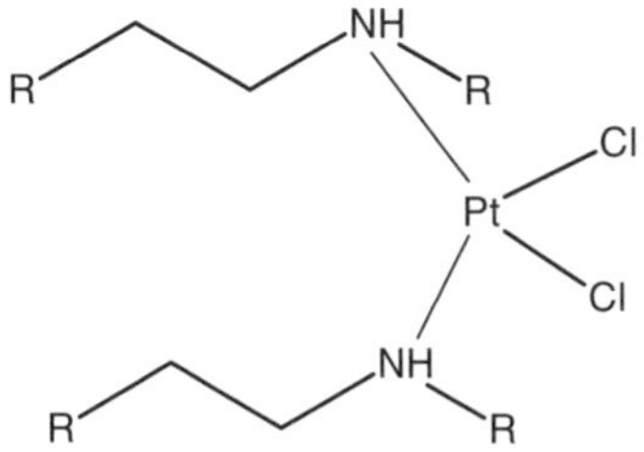

26

It is also possible that only partial reaction occurs giving monoligated structures as **27**.

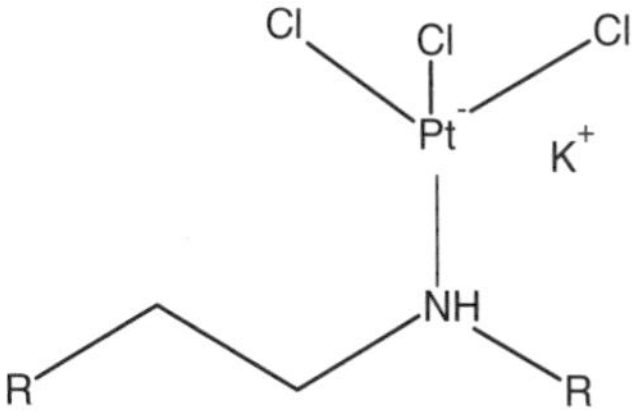

27

Elemental analysis of the PEI/Pt polymer is consistent with a 1:2 ratio of Pt to Cl.

The product contains both DMSO-soluble and DMSO-insoluble fractions. It is reasonable to conclude that the insoluble portion contains at least some crosslinking. Interestingly, the fraction of soluble product increases at high ratios of

K_2PtCl_4 to PEI. Unreacted PEI units are also present. As the ratio of the tetra-chloroplatinate is increased the percentage of platinum-containing moieties also increases.

The products were tested for activity against both L929 and HeLa cells. In such tests, reference compounds should be also be used and the testing should be done against the specific strains used in the overall testing. For cisplatin itself, little to 50% inhibition was found for HeLa and L929 cells beginning at 50 μg/mL. Equivalent activity was found for the polymers at 20 μg/mL for L929 cells and at 6 μg/mg for HeLa cells. Thus, the polymer/Pt complexes showed good inhibition of the cell lines at lower concentrations in comparison to cisplatin itself. The poor result for cisplatin is probably due to reaction with the solvent.

Of interest, these products were tested for bacterial inhibition using a wide range of bacteria. As solids, they were found to be inactive. The DMSO-soluble fractions generally exhibited slight inhibition.

In an attempt to produce water soluble products, a similar study was carried out except using a copolymer of polyvinylamine and vinylsulfonate.[118,119] The copolymer was a random mixture of vinylamine (60 mol%) and vinylsulfonate (40 mol%) units. Again, the reaction was studied as a function of ratio of reactants. Product yield was high for all systems. The ratios of Cl to Pt varied from the expected 2.0 to as high as 2.3 with most about the expected value. This is consistent with a high ratio of the desired dichelation of each Pt atom by two nitrogen sites. The existence of Cl:Pt ratios greater than 2 is consistent with the presence of some units containing $PtCl_3$ complexed with one amine. The loading was also high. The product was soluble in water and DMSO consistent with a minimum of crosslinking. For high ratios of K_2PtCl_4 to copolymer, there was a slight increase in molecular weight from about 4.5×10^4 to 6.5×10^4 with the amount of platinum increasing only modestly. This is consistent with mild crosslinking occurring sufficiently to increase the chain length modestly, but not enough to give insoluble product. A sample unit is depicted below as structure **28**.

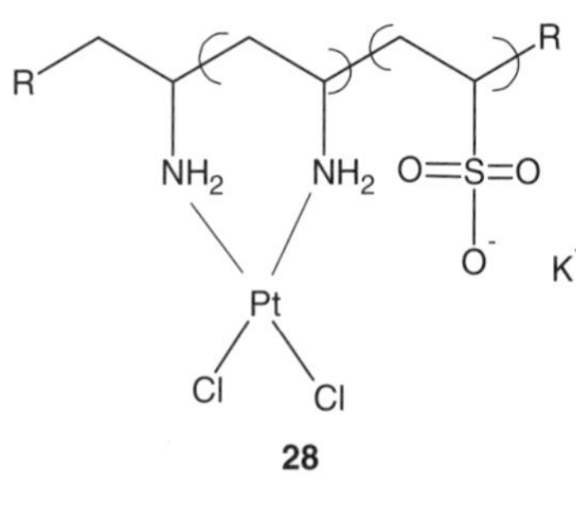

28

The transition between the so called slowly and readily biofissionable products is now discussed. Chitosan was employed in another study attempting to create water-soluble products that were biodegradable.[120,121] Chitosan is a polysaccharide with a structure similar to that of cellulose. Compound **29**, is an abbreviated structure of some of the possible unit structures that include inter/intra-molecularly "di-bonded" platinum products. These structures are similar to the *cis*-DDP structure (structures **A** and **B** below), the monobonded product (structure **C**), and finally unreacted units (structure **D**). Depending on the particular reaction conditions, the resulting product probably contains varying ratios of all four units.

```
           A              B       C       D
         C-O-C     C-O-C   C-O-C   C-O-C   C-O-C
        \   /     \   /   \   /   \   /   \
         C-C       C-C     C-C     C-C     C-C
          |         |       |       |       |
        H-N-H     H-N-H   H-N-H   H-N-H   H-N-H
            \     /         |       |
             Pt           Cl-Pt-Cl Cl-Pt-Cl
            / \             |       |
          Cl   Cl         H-N-H    Cl
                            |
                           C-C
                          /   \
                       C-O-C   C-O-C
                            29
```

The reaction of chitosan and $PtCl_4^{2-}$ was studied as a function of reactant ratio, time, and pH. In general product yields were approximately 50%, based on a dichelated product and percentage of platinum found. Structure **A** represents the preferred structure containing two labile Pt–Cl sites. Structure **B** also contains two labile sites in the cis position, but represents a crosslink or internally cyclized structure, both giving a product with decreased solubility. The amount of nonadjacent same-chain chelation should be low since the chitosan chain is a fairly rigid helix in solution, limiting the close association of different parts of the same chitosan chain with itself.

Solubility was studied in a number of liquids. The lack of complete solubility is consistent with the product containing at least some crosslinks. The molecular weight was determined for the soluble portions. Chitosan has a reported molecular weight (for the sample employed in these studies) of 70,000. The molecular weight of the employed chitosan was determined by light-scattering photometry to be about 60,000, in near agreement with that reported by the manufacturer. This corresponds to an average chain containing about 380 hexose units.

For illustrative purposes we will focus on one product synthesized at a pH of 5–6 using a 1:1 molar ratio of chitosan units to tetrachloroplatinate. This product gives a molecular weight of 1×10^6 after being left in DMSO for 4 weeks. This result is consistent with the following calculations. A chain fully substituted with Cl–Pt–Cl would have a molecular weight of about 111,000. The number of chains that would have to be connected through crosslinking is calculated by dividing the molecular weight found by the molecular weight per chain and is about 9. Elemental analysis shows that the chain is not fully substituted, but only about 35% substituted with Pt functions. Then the new chain would have a calculated molecular weight of 78,000, and the number of chains connected would be about 13.

The amount of crosslinking needed to make a polymer insoluble varies but is generally in the range of 1–5% crosslinking per unit or 1–5 per 100 units. Further calculations can be made that allow an impression of the number of crosslinks and number of Cl–Pt–Cl units per chain. Again, assuming only a 35% reaction (substitution) there would be 66 Cl–Pt–Cl units per uncrosslinked or original chain. Only about two to three of these platinum units need to be involved in crosslinking assuming that only 1% crosslinking is present. This indicates that the vast majority, about 95%, of the platinum sites, are bonded to a single chitosan chin rather than acting as a crosslink.

Infrared spectra of the chelation products prepared using different molar ratios and different pHs are similar. The spectra contain bands characteristic of the presence of both reactants and the presence of new bands consistent with the formation of the Pt–N moiety. The number of Pt–N stretching bands is often taken as evidence of the geometry of the product. Trans-platinum diamines exhibit only one Pt–N band while the cis products exhibit two bands. The second band is weak and sometimes missed. The presence of two bands indicates the cis geometry, but presence of only one band is not firm evidence for the trans geometry. The products contain two Pt–N-associated stretching bands, one at 564 [all bands given in reciprocal centimeters (cm^{-1})] and one at 497, consistent with the cis conformation. A strong band centered around 3430 is assigned to O–H stretching. It remains unchanged in shape, location, and intensity from chitosan itself. The band around 3000 is assigned to N–H deformation. The presence of bands characteristic of O–H stretching and the presence of new Pt–N bands are consistent with reaction occurring through the amines and not the hydroxyl groups. The bands about 1071 are assigned to the C–O stretch in chitosan (both within the hexose ring and the connective ether or acetal linkage). Mass spectroscopy was carried out on the products. The results for the various products are similar.

(1) The percentage chelation ranged from 7 to 43%. Partial chelation is expected because the product precipitates from the reaction mixture, isolating unreacted sites from a ready source of unreacted tetrachloroplatinate(II); and (2) steric factors generally restrict complete substitution on polymers. So far, preliminary test results have shown these materials to not be particularly active in the inhibition of cell lines.

C. Biofissionable Pt–N Complexes Anchored through Primary and Secondary Amines

The Pt–N polymer–metal interconnection in the structures covered in the previous section is generally tight-binding and not readily biofissionable. Therefore, in order to achieve faster release of the monomeric bioactive drug complex in the biological environment, the amino groups must be flanked or carrier-interconnected by bonds that are biocleavable and allow hydrolytic or enzymatic backbone or sidechain fragmentation with ultimate liberation of the monomeric complex entity.

The literature describes an early investigation which successfully probes the feasibility of platinum anchoring via the 1,2-diamine ligand system. Exploiting the ability of polyvinylpyrrolidone (PVP) and a related polymer structure to form molecular complexes with aromatic ring structures containing polar substituents, Howell and Walles[122–125] prepared variously substituted *cis*-dichloro-*ortho*-phenylenediamineplatinum(II) complexes. The common platination method from tetrachloroplatinate(II) anion and diamine in aqueous solution was used. This method allowed the reagents to react with the polymeric carriers in aqueous–aprotic solvents. These products were water-soluble in the lower molecular mass range, and activity was observed in bacterial screens. The following scheme exemplifies the synthetic sequence for the 4-hydroxy-substituted aromatic reactant. In these molecular complexes, release of the bioactive component obviously does not require a covalent bond-breaking step, but merely a dissociation of the complex (**30**) that is loosely held together through hydrogen bonding.

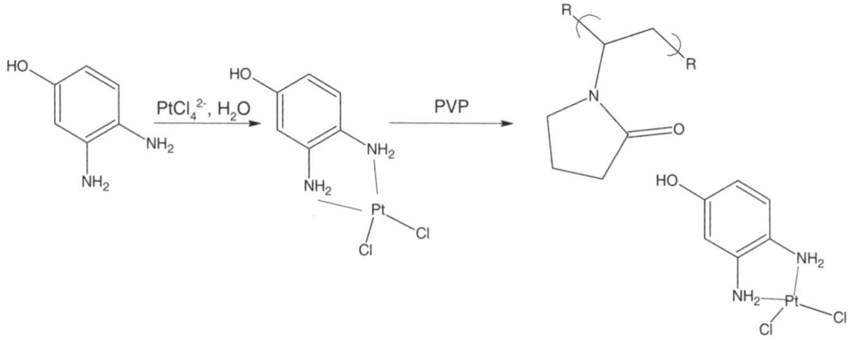

30

Many ringed systems are included in the Howell–Walles patent,[125] including a wide range of nitrogen-containing systems such as piperidones, oxazinidinone, oxazolidinone, morpholinone, caprolactam, succinimid, imidazolidinone, cyanuric acid, and hydantoin.

Along with such complexes, their patent included a wide variety of compounds that contain platinum bound through adjoining aromatic amines. Many of these are based on substituted o-phenylenediamine where the substituting group, R_1, can be an acid, alcohol, sulfonic, or other compound that can subsequently be connected to a polymer through a condensation or addition reaction of the complex (**31**).

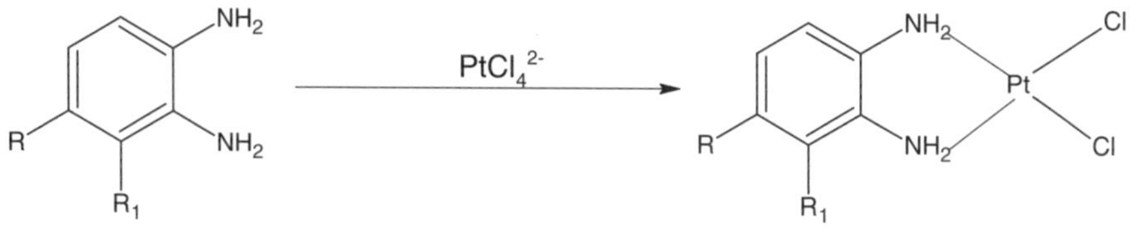

31

In Howell's work the diamine ligand system was generally aromatic in nature. In contrast, conjugates featuring metal coordination by an aliphatic 1,2-diamine segment were disclosed in a patent by Arnon et al.[126] These authors treated ethylenediamine-modified dextran polymer, dissolved in water, with tetrachloroplatinate(II) dianion and obtained platinum-containing conjugants. Significant in vivo antineoplastic activity was reported.

A major program aimed at the platination of carriers used was initiated in the late 1980s in the laboratory of Eberhard Neuse.[127–142] In that still ongoing program, emphasis has been on synthetic macromolecules as the preferred carrier constructions. Use of manmade polymers provides an advantage insofar as their synthesis permits tailoring and fine-tuning of both main chain and side-group structures in accordance with critical biomedical specifications. (See Section V.) In addition, properly constructed synthetic macromolecules are generally less immunogenic than natural polymers, notably those based on a proteinaceous composition. Finally, they can be designed so as to remain reasonably stable while in central circulation, yet degradable on entering intracellular space. This is a clear advantage over non-degradable carriers used in many studies.

The polymers of choice in this program are of the polyaspartamide type, readily obtainable by the unusually versatile technique of stepwise aminolytic treatment of polysuccinimide pioneered in a Czech laboratory.[143,144] This synthetic approach is illustrated below (there is a random distribution of subunits along the polymer chain, **32**). This methodology allows successive imide ring opening by amine nucleophiles R–NH$_2$ with conversion to aspartamide units and permits the preparation of poly-α,β-DL-(aspartamides) equipped with two or more different side functionalities R.[140] Aqueous dialysis in membrane tubing with 25,000 molecular mass cutoff provides the product polymers in a molecular size range substantially above the kidney threshold.

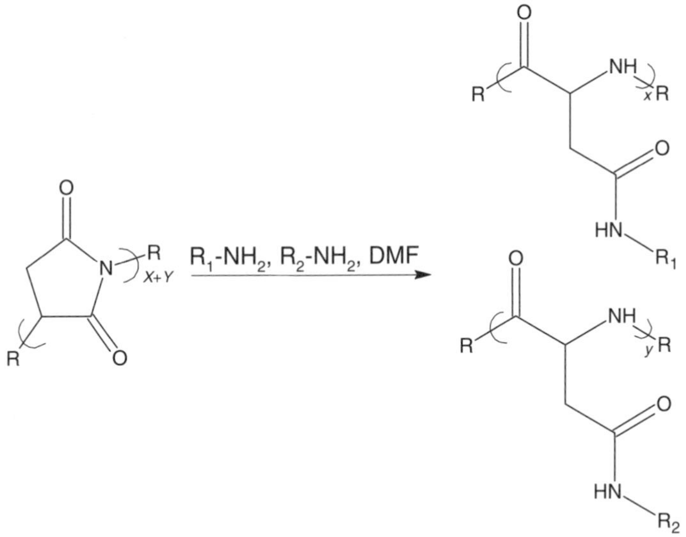

32

The water-soluble macromolecules of this type contain both D– and L–configurated methyl carbon in the backbone, as well as α- and β-peptidic bonding schemes. (For simplicity, only the α form is shown here and in subsequent illustrations.) While the polyaspartamides are susceptible to degradation by endo- and exopeptidases for ultimate catabolic elimination, the sole end product of mainchain fragmentation is aspartic acid, a nontoxic and body-friendly catabolite. Furthermore, the presence of D-configured β-peptidic units prevents rapid enzymatic fragmentation ("unzipping"). As a consequence, the degradation rate is quite slow and allows substantial mainchain survival while the assembly is still in serum circulation.

The process given above has been utilized extensively for the synthesis of water-soluble carriers in which R_1 contains a hydrosolubilizing functionality. Such R_1 functions as hydroxyl or *tert*-amine side-group terminals are the majority components in the overall structure (**32** top/**32** bottom = 3–9) and they ensure adequate water solubility. Oligo(ethylene oxide) chains have also been introduced as R_1 substituents aiding in solubility and acting as protectors against undue serum protein binding. The group R_2 serves to bind the platinum atom. Carriers designed to be platinated via 1,2-diamine coordination constitutes an ethylenediamine (*en*) segment. The reaction exemplifying metal anchoring to a representative ethylenediamine-functionalized carrier synthesized is depicted below (**33**), where R_1 = Me$_2$N–CH$_2$CH$_2$CH$_2$– and R_2 = H$_2$N–CH$_2$CH$_2$–NH–CH$_2$CH$_2$–) and also described in **34**.

Platination is smoothly brought about in aqueous solution, with potassium tetrachloroplatinate used as the platination agent and mild solution acidity maintained

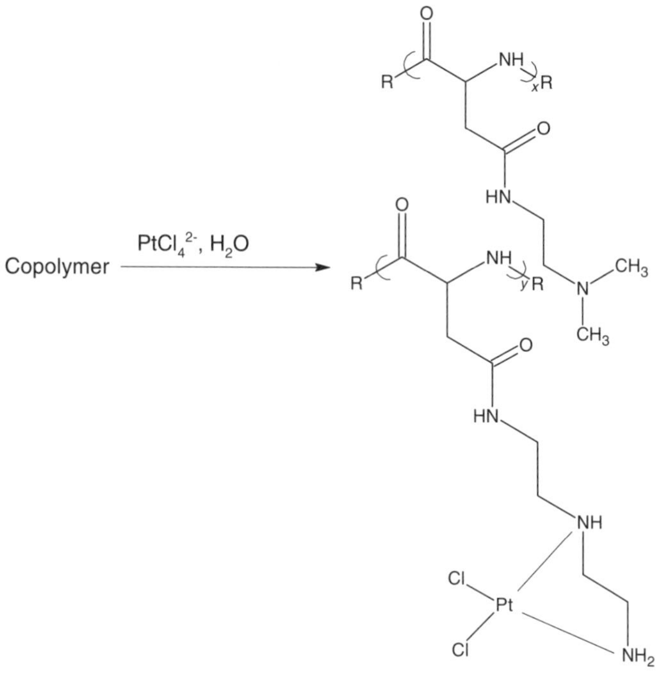

33

at pH 5–6. After aqueous dialysis in the presence of chloride anion, the conjugates are isolated by freeze-drying of the retentate as water-soluble solids typically containing 10–20% Pt by mass. Pt/R_2 molar ratios of 1.15–1.2 in the reaction system were found in the early work to be adequate for substantially quantitative metal incorporation. This is still the adopted range in current work involving differently structured carriers. For the platination of polymers containing basic hydrosolublizing units, such as a tertiary amine functionality as in a sequence such as pictured in **32**, increasing the platinum feed above 1.2 equiv must be avoided, as this would cause platinum loading to exceed the maximum dictated by the proportion of R_2 in the polymer. Such excess loading presumably involves ionic bonding through salt formation between $[HPtCl_4]^-$ anion and the protonated tertiary amino groups.

For platinum incorporation through intrachain-type *en* segments, Scheme 2b, the basic platination procedure is the same as used for backbone, Scheme 2a, reactions. However, an increased Pt/en feed ratio was found to be required for complete loading probably due to steric factors. The range of 1.23–1.3 was satisfactory for most of the carriers investigated. Structure **34**, shows an exemplifying reaction involving an oligo(ethylene oxide)-modified carrier prepared by interfacial polymerization.

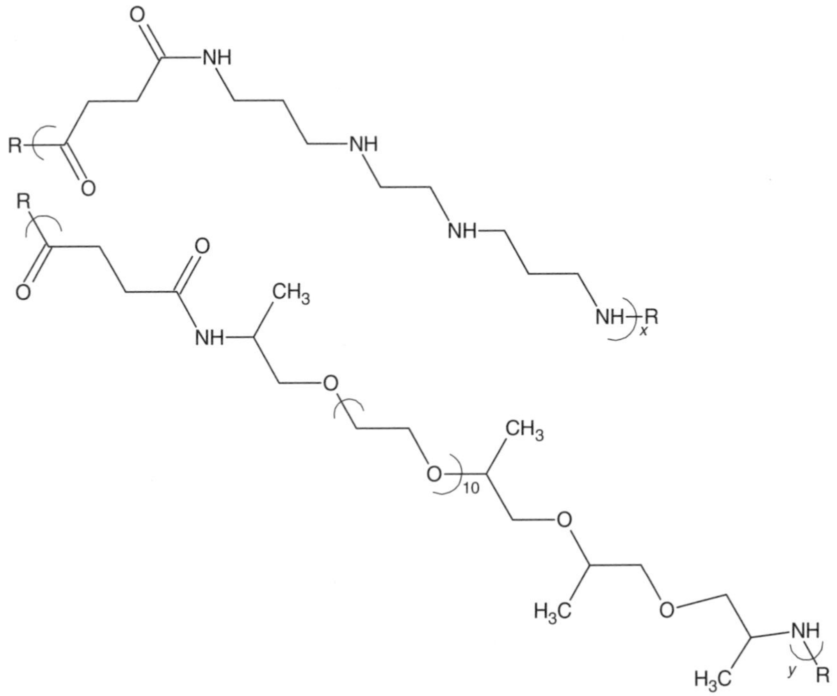

34

After reaction with the tetrachloroplatinate, structure **35** forms only from the amine-containing units as shown in that structure **(35)**.

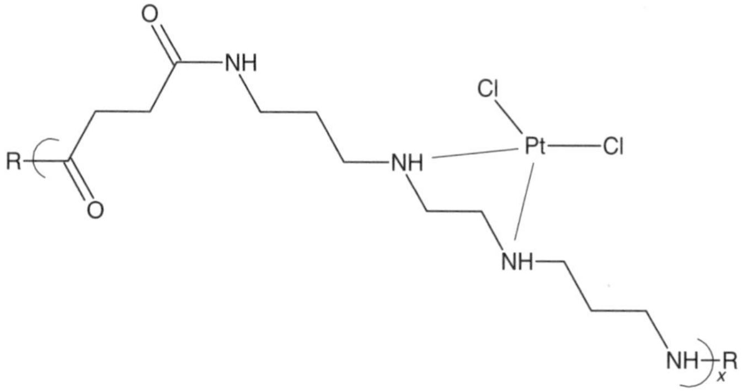

35

While the 6-membered platinum-containing unit can also form, the majority of the product will carry the 5-membered platinum ring as shown.

The conjugates so far presented in this section are characterized by an anchoring mechanism involving ethylenediamine binding to the metal atom, resulting in the generation of a cisplatin-type complex with *cis*-diamine-coordinated platinum. The polymer anchoring may also involve metal binding through a single amine ligand. Here, monoamine-functionalized carriers, typically of the polyaspartamide type, are used. The product polymers, arising via anionic intermediates, are presumed to be essentially neutral structures as shown in Scheme 3. A representative platination step conducted with a Pt/NH$_2$ molar feed ratio of 1.4 and, again, with tetrachloroplatinate(II) metallation agent, is depicted in, **36**.

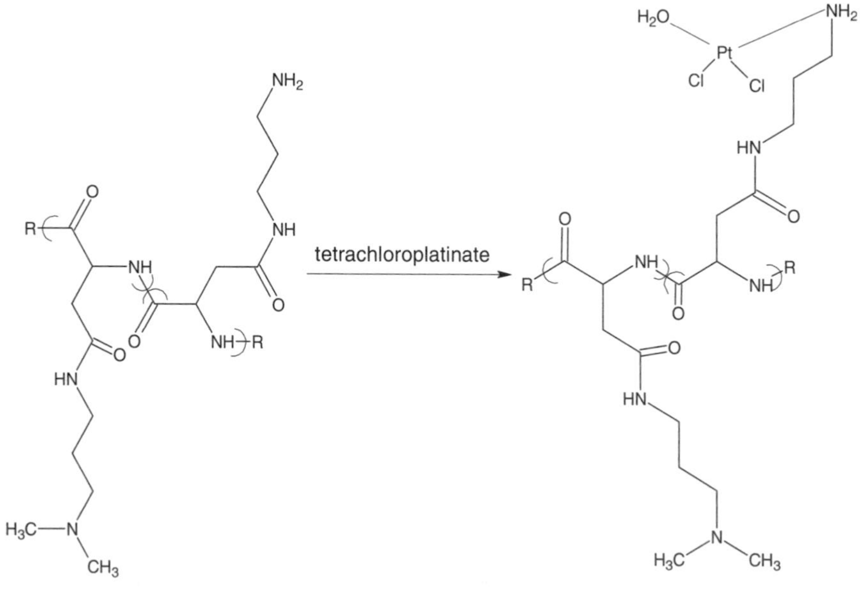

36

An excess of the NMe_2 terminated copolymer unit is employed in this system. Again, while all the nitrogen-containing moieties can react with the tetrachloroplatinate, the most active is the terminal nonsubstituted amine.

The IC_{50} value, which represents the drug concentration required to inhibit cell growth by 50% relative to a drug-free control, will be used to compare polymer systems. The 1,2-diamine-coordianted platinum polymers [types (A) and (B), Scheme 2] were screened in vitro for antiproliferative activity against HeLa cells. All conjugates tested were found to be active. Interestingly, the conjugate of Scheme 2, representing type (a) structures, performed best, with $IC_{50} = 14$ µg Pt/mL. In contrast, type (b) conjugates, all characterized by the presence of oligo(ethylene oxide) segments in the backbone, gave IC_{50} values some 3–4 times larger, indicating considerably lower activities. This performance difference may simply be a consequence of the known propensity of poly(ethylene oxide) segments to block access by extraneous proteins, thus inhibiting swift cleavage of the biofissionable sites by peptidases and other proteases. This retarded fragmentation may, on the other hand, well be of advantage in in vivo administration.

Another observation, as yet open to interpretation, concerns the potential biological function of the hydrosolubilizing unit R_1 in **32**. If the 3- (dimethylamino)propyl (DP) substituent in the conjugate of **33** is replaced by the 3- (morpholin-4-yl)propyl (MP) group, the antiproliferative activity is grossly reduced ($IC_{50} = 42$ µg Pt/mL). Furthermore, replacement by the 2-(2-hydroxyethoxy)ethyl side groups leads to an even more dramatic activity reduction ($IC_{50} > 80$ µg Pt/mL). Decreasing activity going from a DP- to an MP-modified carrier is evidenced also in a study of conjugates featuring monoamine-coordinated platinum (Scheme 3). When tested for antiproliferative behavior toward the LNCaP human metastatic prostate adenocarcinoma cell line, the DP-containing conjugate of **33**, and a similar DP- substituted relative, both show a surprisingly good performance ($IC_{50} = 5$ µg Pt/mL). The corresponding MP-modified conjugates exhibit IC_{50} values more than 5 times higher, reflecting considerably diminished activities. A similar trend is even more convincingly apparent in a test series against HeLa cells, where various DP-containing polymers show excellent cytotoxic behavior. IC_{50} values were in the range of 0.5–3 µg Pt/mL, while compounding polymers with $R_1 = 2$-hydroxyethyl are practically inactive within reasonable concentration ranges. These observations serve to emphasize the critical importance of the hydrosolubilizing side groups in controlling the conjugates' pharmacokinetics.

Duncan, et al.[145–147] have investigated a number of copolymers having a backbone based on *N*-(2-hydroxpropyl)methacrylamide (HPMA, **37**). Briefly, sidechains containing various amino acid (peptidyl) spacers are introduced onto the HPMA polymer. These peptidyl sidechains include Gly–Gly and Gly–Phe–Leu–Gly that contain carboxylate or amino end-groups that are reacted with cisplatin.

The diglycyl spacer was used because it is nonbiodegradable, whereas the tetrapeptide spacer is known to be cleaved by the lysosomal thiol-dependent proteases. The platinum loading is ~3–7 wt%. In vitro, the HPMA copolymer platinates show a range of platinum release rates at pH 7.4 and 5.5 ranging from less than 5% for 24 h for the diamino species that requires enzymatic activation, to greater than 80% in 24 h for the carboxylate product. (The topic of carboxylate-bound platinum products is dealt with in greater extent in the following section.) Cisplatin and the fast releasing carboxylate copolymer showed IC_{50} values of 10 micrograms/mL Pt equivalent against B16F10 cells in vitro, whereas the slowly releasing copolymers were not cytotoxic over the dose range studied.

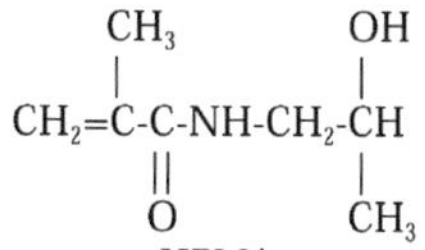

$$CH_2=C-C-NH-CH_2-CH$$

CH₃ ... OH

HPMA

37

The copolymers were studied using L1210 and B16F10 tumors inoculated intraperitoneally (IP). Neither cisplatin nor the HPMA copolymers were active against IP. B16F10 tumors when administered IP. But when the platinum-containing copolymers were administered intravenously (IV) to treat subcutaneous (SC) B16F10 tumors grown to palpable size, free cisplatin was still inactive but the HPMA copolymers showed significant antitumor activity.

The platinum-containing copolymers were 5–15-fold less toxic than cisplatin in vivo. After IV administration, the blood clearance of the copolymers was slower ($t_{1/2}$ ~10 h) than that of cisplatin itself ($t_{1/2}$ < 5 min). The platinum- containing copolymers showed a 60-fold increase in Pt accumulation in the B16F10 tumor tissue in comparison to cisplatin itself.

Such studies illustrate the potential for administrating larger dosages of platinum but that the anticancer activity of polymers is complex.

X. Pt–O-BOUND POLYMERS

The incorporation of platinum through carrier-attached amine ligands provides a tight-binding system. An alternative strategy is based on anchoring structures using carrier-bound ligands of the leaving-group type, employing the more labile Pt–O linkages to attach the platinum-containing moiety. A number of these employ dichloro[rel-1*R*,2*S*]-1,2-cyclohexanediamine-N,N′-platinum(II), DACH-Pt, as the platinum reactant. DACH-Pt **(38)** is made through the reaction of DACH with tetrachloroplatinate. The DACH-Pt compound is given several names, but the most widely used ones are *trans*-1,2-diaminocyclohexanechloratoplatinum(II), *trans*-1,2-diaminocyclohexane Pt(II) (without specification of the particular salt), and *trans-cis*-1,2-diaminocyclohexane platinum(II).

Numerous related compounds have been synthesized and used in place of the dichloride. These include the dibromo, oxalato, malonato, dinitrato, sulfato, mono, and bis(D-glucuronato) Pt(II) compounds.[148] A number of these compounds were found to inhibit L1210 mouse leukemia cells; the oxalato, malonato, dinitrato, and mono(D-glucuronato) Pt(II) compounds were most effective.[148]

A related early study of dichloro, oxalate, malonate, methylmalonate, and the uracil complex also showed antitumor activity.[149] They investigated the *cis, d-trans,* and *l-trans* complexes and found that the *l-trans* complexes showed the greatest neoplasm inhibiting activity. Therefore, much of the effort has concentrated on this particular isomer.

Such studies spurred on investigations involving *trans*-1,2-diaminocyclohexane Pt(II) complexes including a number of polymeric materials.

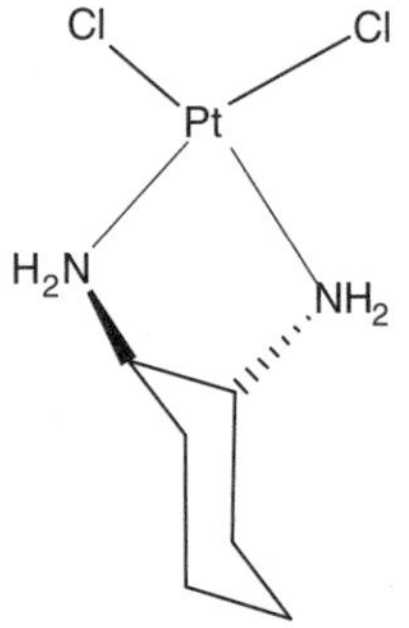

38 (DACH-Pt)

Carboxyl (Scheme 4 showing the dicarboxyl compound) or hydroxyl (Scheme 5 showing the dihydroxyl compound) groups are preeminently suitable functionalities for this anchoring mode.

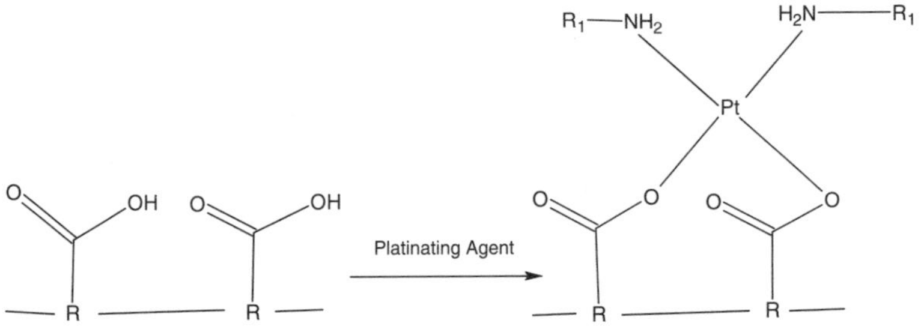

Scheme 4 Formation of platinum-containing compounds from reaction with dicarboxyl-containing compounds.

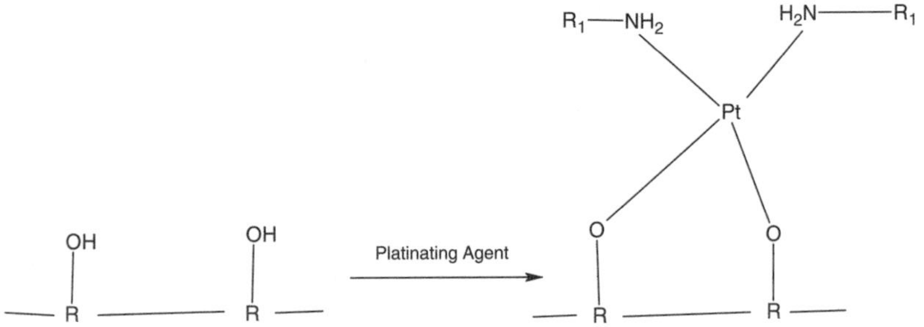

Scheme 5 Formation of platinum-containing compounds from reaction with hydroxyl-containing compounds.

Here the polymer–metal connection, of the comparatively weak carboxylato-Pt, or of the even more labile hydroxylato-Pt types, is susceptible to hydrolytic cleavage, notably in the acidic milieu (pH ~ 5) maintained in the lysosomal compartment. The diaminoplatinum part of the complex is thereby released from the carrier without the need for other biofissionable linking groups. Aqua ligands take the place of the carrier-attached leaving groups in this process. Release kinetics, hence, are quite different from those involving the tight-binding anchoring discussed in Sections VIII and IX. The coordination pattern is preferentially designed so as to contain two carboxylato (or hydroxylato) ligands in cis geometry chelating the metal atom. These are in the form of both kinetically and thermodynamically stabilized 6- or 7-membered ring structures. Chelation is important here as the enhanced stabilization will serve to protect the ligand–metal bonds against premature hydrolysis while the conjugate is still in serum circulation.

In one of the first studies using the carboxylato-anchoring concept, Holtzner and collaborators[150] used dinitrate or diaqua derivatives of the cis-diamineplatinum(II) complex system as the platination agents, anchored the metal to carboxyl functionalities of poly(glutamic acid), poly(aspartic acid), or poly(L-lysine). The reactions were conducted in aqueous phase. The resulting conjugates were purified by dialysis but not isolated. They served nonmedical purposes, acting as spin label carriers in physical polymer characterization studies. Notwithstanding their different applications, these conjugates may well be seen to be the forerunners of the numerous subsequent investigations utilizing the carboxylato-Pt anchoring strategy for chemotherapeutic developments.

A project under study since the mid-1980s by a group at the Weizmann Institute of Science[151–166] stands out for its wide scope and proficiency. The project has been focused on the synthesis and biomedical evaluation of polymer–platinum conjugates antibody-modified for immunotargeting to human tumors. In early work, these researchers treated a variety of carboxyl-functionalized polymers, such as carboxymethyldextran, divinyl ether–maleic anhydride copolymer, poly(glutamic acid), and poly(aspartic acid), with cisplatin or its aquated derivative, followed by dialysis for removal of uncomplexed platinum compounds. The product conjugates, believed to feature the anchoring patterns (a) and (b) in Scheme 2 using general compounds as given in Scheme 4, exhibited reduced serum protein binding and extended circulation half-lives. Most of them were quite active, both in vitro and in vivo. They showed a wide therapeutic range of activity in animal tests because of low toxicity. Conjugates based on poly(glutamic acid) and poly(aspartic acid) were notably efficacious, exceeding cisplatin efficacy. Among the carboxymethyldextran-based polymers, the type having a carrier molecular mass of 40,000, in contrast to those with considerably lower or higher chain lengths, was also identified as a top performer. Representative carriers of that type were then, with the aid of a carbodiimide coupling agent, connected to an antibody specific to murine B lymphoma cells. On chromatographic purification, the conjugates were platinated in aqueous solution. The antibody-modified conjugates, with up to 50 mol of Pt complex per mole of antibody, were found to be preferentially cytotoxic to the substrate cells in comparison with free carrier, platination agent, or conjugates comprising nonspecific antibody.

Building on these earlier findings, the group has more recently tested two representative conjugates, a carboxymethyldextran-platinum copolymer and a platinated poly(glutamic acid), both in vitro and in vivo against human ovarian carcinoma. The LD_{50} values of 9 and 16 μg Pt/mL, obtained for the two conjugates in cell culture tests against the IGROV-1 line, are only slightly larger than determined for cisplatin (6 μg Pt/mL). In view of the lower toxicity of these polymers, they exhibit a high efficacy. Even more convincing are the results of in vivo tests against xenografted OVCAR-3, giving 1.4–2.7-fold longer survival times, which are superior to cisplatin.

Widening the scope of exploration, Weizmann scientists have more recently embarked on a program exploiting the benefits of the avidin–biotin coupling and homing strategy so successfully utilized in other biomedical projects.[166] In their quest for more efficacious immunotargeting of platinum conjugates, they have developed a two-component system typically consisting of a biotinylated antibody and an avidin- or streptavidin-modified carboxymethyldextran-platinum conjugate. (For details, see cited references.) The components are mixed in solution just prior to administration in murine tests (to avoid premature interaction and crosslinking); alternatively, they are administered in a sequential fashion. This strategy has been successfully tried, for example, with a specific monoclonal antibody (m Ab 108) in the treatment of a xenografted human epidermoid carcinoma. In a modified form, it could be useful also for liver accumulation of polymeric anticancer drugs, radionuclides, and other agents. Using a trinitrophenyl-substituted steptavidin, the researchers coupled this hepatic targeting component to a biotinylated carboxylmethyldextran, followed by platination with cisplatin. In contrast to free cisplatin, this macromolecular complex selectively targets the liver (specifically the Kupffer cells), in which the active platinum compound persists for 9–15 h and longer.

The biotin-avidin technique as practiced by Schechter's group for immunotargeting offers outstanding promise for the utilization of monoclonal antibodies in cancer chemotherapy.

Several other research groups, taking advantage of existing or preparatively introduced carboxyl functionality in polysaccharides, have used modified carbohydrate polymers as carriers for platinum complexes. Thus, Fujii and Imai with colleagues[167–170] reported significant pharmacokinetic benefits resulting from complexation of cisplatin with sodium alginate with molecular masses of 9000 and 40,000. Conjugation was achieved by thorough shaking of these polymers with cisplatin in water in a 1:10 molar ratio of cisplatin to uronic acid residue, followed by chromatography on Sephadex G25. The resulting solutions have at least 90% of the metal in complexed form [presumably as in type (a), Scheme 2]. They display in vitro anticancer activity comparable with that of free cisplatin, coupled with lessened dissipation into other organs and retarded elimination from plasma. The latter is most pronounced with the alginate of molecular mass 40,000.

In a related study in Duncan's laboratory[171–178], carboxymethyldextran has been platinated with cisplatin, achieving metal coordination involving carboxyl groups of two adjacent sugar units. Tests in vitro against the L132 human embryonic lung carcinoma cell line reveal cytotoxic activity almost as high (IC^{50} = 5.47 μg Pt/mL) as shown by cisplatin (3.91 μg Pt/mL), although in vivo tests point to considerably lower survival rates. The same group has reported conjugates based on alginates, poly(glutamic acid), and a *N*-(2-hydroxypropyl)methacrylamide (HPMA) copolymer. The carboxymethyldextran conjugates perform as well as free cisplatin [in contrast, e.g., to poly(glutamic acid) conjugates with activities more than 10 times lower]. Like other biopolymer-based conjugates, they become insoluble on short-term storage, a deficiency not shown by the HPMA-derived polymers.

Depending on the particular polymer, the platinum is attached through Pt–N, and Pt–O and probably through other sites. The tetrachloroplatinate is the preferred reactant for Pt–N complexes such as **39** (and of the form presented in Section VIII), while cisplatin itself is used in the formation of Pt–O-bonded complexes **(40)**.

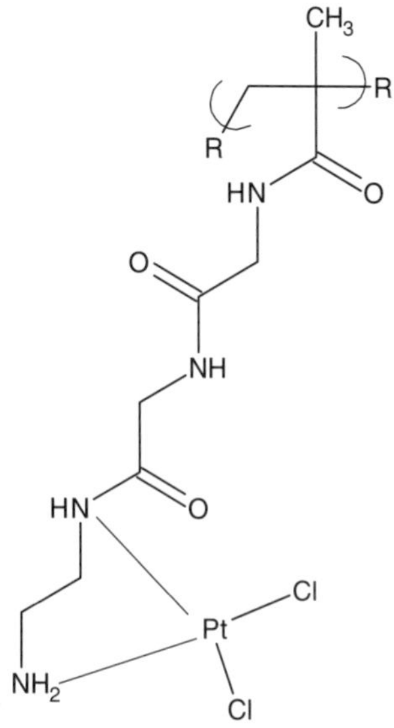

39

Duncon's group has also included platinum into dendrites. The dendritic polymer platinates exhibit high drug efficiency, high drug carrying capacity, good water solubility, good stability on storage, reduced toxicity, and improved antitumor activity in vivo. For example, polyamidoamine Starburst (PAMAM) dendrimer generation 3.5 was reacted with cisplatin in a cisplatinum to dendrimer molar ratio of 35–1 giving a product with about 25 wt% platinum. Other dendrimer families were also prepared including polypropyleneimine dendrimers with either a diaminobutane (DAB, Astromol) or a diaminoethane (DAE) core. The PAMAM-Pt dendimer, derived from reaction with cisplatin, was active IP. against L1210 and i-b. B16F10 tumors in vivo whereas cisplatin itself was not. The dendrimer-Pt was also less toxic (3–15-fold) than cisplatin in animal studies.

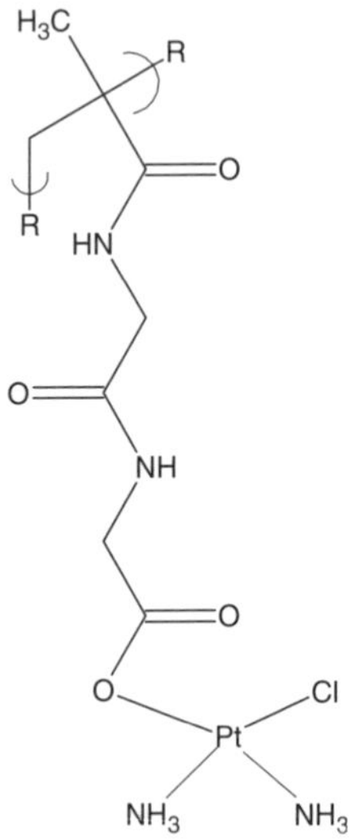

40

Functionalized poly(amido amine)s were also made by Duncan's group. These materials were often coupled with cyclodextrin. Structure **41**, contains one of the poly(amido amine) carriers. Here R_1 is β-cyclodextrin. The reactive platinum- containing species is cisplatin itself so that the platinum-containing sites are similar to those pictured in **40**. The products showed in vitro cytotoxicities lower than those of cisplatin. Some of the products were more toxic than cisplatin against L132 cells. It is believed that some of the products are neither cytotoxic nor haemolytic, so the mechanism of the higher antitumor activity is not clear. For instance, typically polymeric antitumor agents of this form show reduced activity compared to the parent cisplatin, when tested in vitro. Their cellular pharmacokinetics are different. Cisplatin enters the cell very rapidly by transmembrane passage. However, the polymer conjugate is internalized more slowly by the mechanism of endocytosis, and thereafter the biologically active platinum species are also liberated relatively slowly.

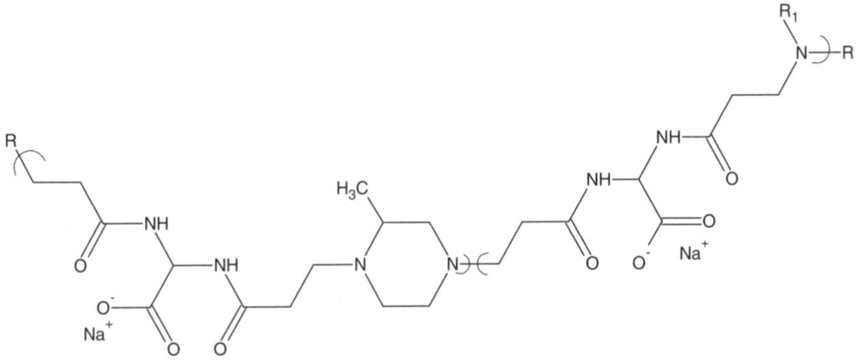

41

The antitumor activity was also tested in vivo looking at i.p. introduced L1210 tumor cells. The maximum tolerated dose, MTD, for cisplatin was 2 mg/kg. This dose the T/C (ratio of mean survival of treated animals to mean survival of controls) was 171%. The Pt complex of **41** showed a T/C similar to that of cisplatin with a lowered toxicity (MTD = 5 mg/kg). The conjugate formed from the first unit in **41** was about threefold less toxic with a T/C of 192%.

Yuichi Ohya and co-workers have employed a number of platinum-delivery agents including dextran and oxidized dextran derivatives, and dicarboxymethylated dextrans[179-189] and poly(ethylene glycol).[190]

Emphasizing the dextran and dextran derivatives effort, carboxymethyldextran was platinated by Ohya's team as similarly described by Schechter, but with trans-1,2-diaminocyclohexanechloratoplatinum(II) in place of cisplatin as the platination agent [type (a), Scheme 2, RR = *trans*-cyclohexane-1,2-diyl]. Furthermore, *cis*-dicarboxylatoplatinum-type conjugates have been derived from periodate-oxidized dextran featuring subunits with the carbohydrate ring opened to expose two carboxyl functionalities in 1,3 geometry. Platinum conjugation of a partial segment structure of the carrier, in accordance with type (b), Scheme 2, is given in structure **42**. A different *cis*-dicarboxylatoplatinum coordination pattern involving 1,1-dicarboxylatometal chelation is realized in conjugates derived from a malonic acid-modified dextran conforming to Scheme 4. In contrast to the platinum polymers thus far discussed in this section, the water-soluble conjugates prepared in Ohya's laboratory have been physically isolated in the solid state. Screens in vitro against p388 lymphocytic leukemia cells show the comparatively labile monocarboxylato-coordinated conjugates to be of low cytotoxic activity. However, the chelate-stabilized dicarboxylatoplatinum-type conjugates have nearly equivalent activities to free unconjugated diaminocyclohexaneplatinum complexes.

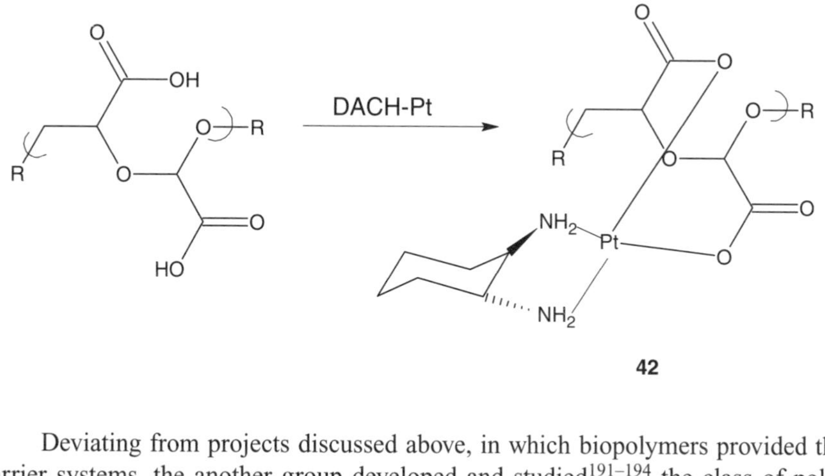

42

Deviating from projects discussed above, in which biopolymers provided the carrier systems, the another group developed and studied[191–194] the class of poly-(ethylene oxide)–poly(aspartic acid) and poly(glutamic acid) block copolymers serving as micelle-forming macromolecules. Water-soluble polymeric micelles offer a special advantage in drug delivery. They can act as drug carriers by incorporating the medicinal agent for protected transport to the target, even if the stabilized inner core is quite hydrophobic. The researchers have now employed these copolymers for platinum incorporation by treatment with cisplatin in aqueous solution. Micelle formation depends critically on the cisplatin/Asp molar feed ratio; it is optimal at a ratio of 1:1. The cisplatin reactant in this preparative step partly or completely exchanges its chloro leaving groups against carrier-bound carboxyl functionalities with generation of both monocarboxylatoplatinum links and intra- or intermolecular dicarboxylatoplatinum bridges. A typical conjugate segment is shown in **43** with Pt coordinated by a monocarboxylato ligand provided by an α-peptidic aspartic acid segment. Cytotoxicity of the conjugates against B16 murine melanoma cells has been reported to be about one-fifth that of free cisplatin.

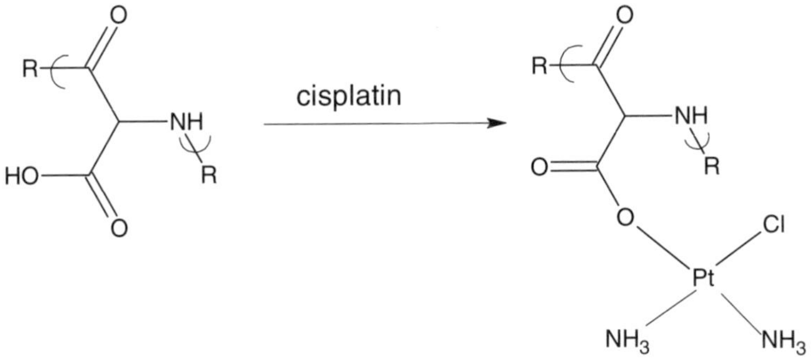

43

A different type of manmade polymer has been used for platination by Bogdanov and collaborators. Along with studying platinum-containing monomer attachments to DNA,[195,196] they investigated a number of copolymer products.[197–199] Poly(L-lysine) can be modified by acylation of ∈-amino side groups with succinic anhydride and partial (~28%) substitution of the carboxyl sidechain terminals generated with poly(ethylene oxide).[199] Treatment of these carriers with cisplatin in aqueous dimethylformamide solution, followed by chromatographic purification, affords water-soluble conjugates (described as adducts by the authors). Pt contents were typically 4.3% by mass. Spectroscopic observations suggest bifunctional binding of the platination agent, largely through intramolecular bridging between two carboxyl groups. However, a subpopulation of about 20% of released Pt is carrier-bound with much lower affinity, indicating a more labile anchoring mechanism. Tested in vitro against a Bt 20 human mammary adenocarcinoma cell line, the conjugate shows the same high cytotoxicity as observed with a PEO-free analog, with IC_{50} typically 0.9 μM Pt as against 0.3 μM Pt for free cisplatin. It follows a pharmacokinetic pathway, however, which is different from that of its PEO-free counterpart. Thus, in contrast to the latter, it experiences long circulation in the bloodstream and significantly increased accumulation in a mammary tumor, yet reduced accumulation in the kidneys.

A Korean laboratory has synthesized and studied a number of platinum-containing monomeric [200–207] and polymeric[208,209] products. Manmade polymers, again, were the basis of the polymer aspect of this effort. Han and coworkers synthesized a series of carriers of the poly(tetrahydropyran-5,6-diyl)(1,2-dicarboxyethylene) type by free-radical polymerization of maleic anhydride with variously substituted dihydropyranes. These inherently bioactive compounds were then platinated with the aquated form of cisplatin. The best performers in cell culture tests, the pyran-5,6-diyl derivative (**44**) and a related pyran-2,6-diyl derivative (**45**), have proved to be excellent candidates for further studies involving platination with aquated cis-di(cyclopropylamine)platinum(II) and DACH-platinum complexes. Molar feed ratios (polymer repeat unit/Pt) of 2:1 result in platinum loading of ~10–14% by mass. Conjugates of type given in Scheme 4, were screened in vitro against several human cancer cell lines and showed high activity. The two platination products, **44** and **45**, for example, were at the top of the range, giving IC_{50} values of about 5µg Pt/mL against the A549 nonsmall cell lung cancer, and about 10 µg Pt/mL against the NCT-15 colon tumor line. Increased lifespan (ILS) data for the two conjugates obtained in in vivo tests against murine L1210 leukemia approach 140% (cisplatin: 180%), and for a related di(cyclopropylamine)platinum polymer ILS is increased to 223%, warranting further investigation.

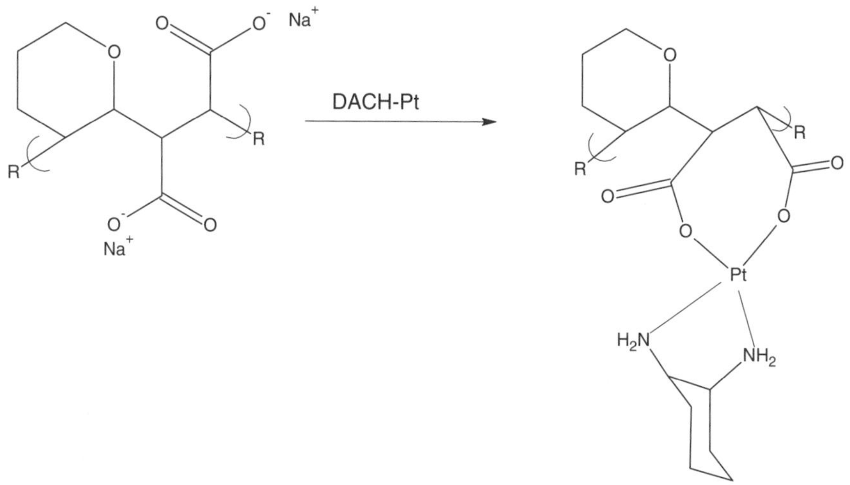

44

The products shown here will also contain unreacted units. Further, the structure shown below will have a combination of platinum-containing structures including the nonpyran platinum-containing structure shown above as well as the structure shown in **45**.

These products are made from the ring-opening reaction of the copolymer formed from divinyl ether and maleic anhydride known as DIVEMA (**45**). DIVEMA, and its polyions, are interferon-inducing agents and possess antitumor activity. This is another illustration of the effective coupling of divergent agents that exhibit antitumor activity.

Wang [210] synthesized a number of similar platinum-containing polymers derived from ring-opened DIVEMA (46). Along with the 1,2-cyclohexanediamine platinum compound, other complexes of cis-PtL_2Cl_2 and cis-PtL_2I_2 were studied. These include compounds where L = NH_3, isopropylamine, aziridine, and L_2 = 2,2'-bipyridine, 1,3-propanediamine, and o-phenylenediamine.

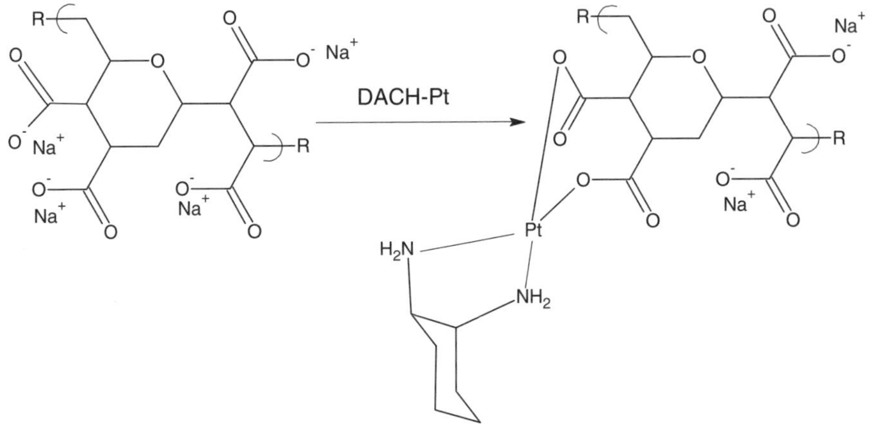

45

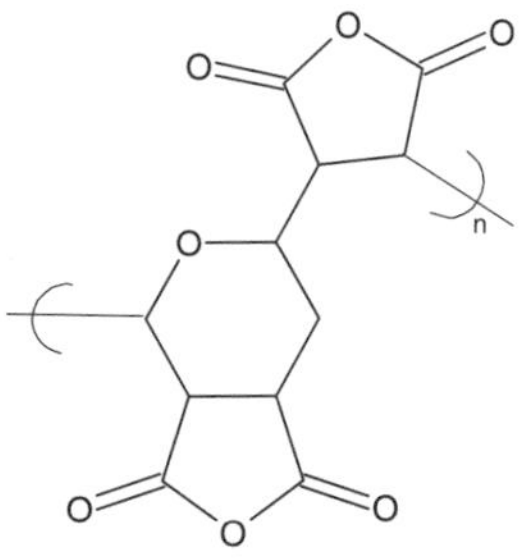

DIVEMA

46

Neuse and his group[211] are also involved in the synthesis of platinum- containing drugs with Pt–O linkages through dicarboxylato ligands. Here, the carboxyl groups required for chelation are arranged pari-wise as either backbond segments or as side-group constituents.

Representative conjugates featuring a 1,1-dicarboxylatoplatinum complex result from aqueous-phase platination of a carrier derived from 2,2-dicarbomethoxymalonic acid and aliphatic diamines by ester-amine polycondensation. Products like **47** are formed which illustrate the reaction. Trans-1,2-diaminocyclohexanediaquaplatinum (III) dinitrate (DACH-Pt) used as the platination agent.

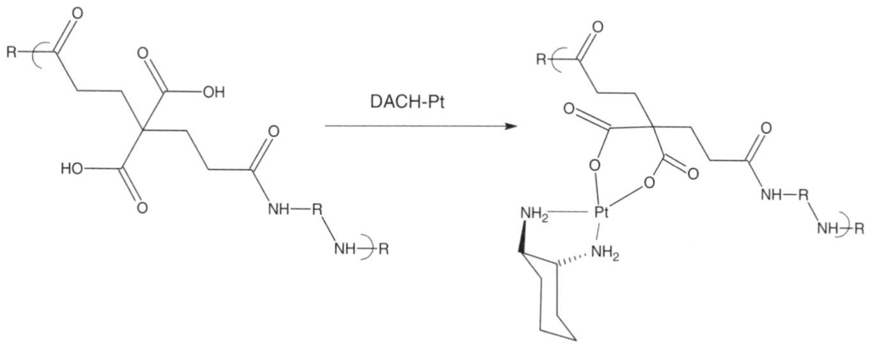

47

where R$=-CH_2-CH_2-CH_2-O-CH_2-CH_2-O-CH_2-CH_2-O-CH_2-CH_2-CH_2-$ or $-CH_2-CH_2-CH_2-$.

Carriers similar to the conjugates shown in **32** were synthesized by Michael addition polymerizations or from polysuccinimide by a two-step ring-opening procedure. These carriers contain 1,2-dicarboxyl-functional anchoring sites in the side groups. They can be platinated to make structures like those in **49**, derived from **48**, and **50**.

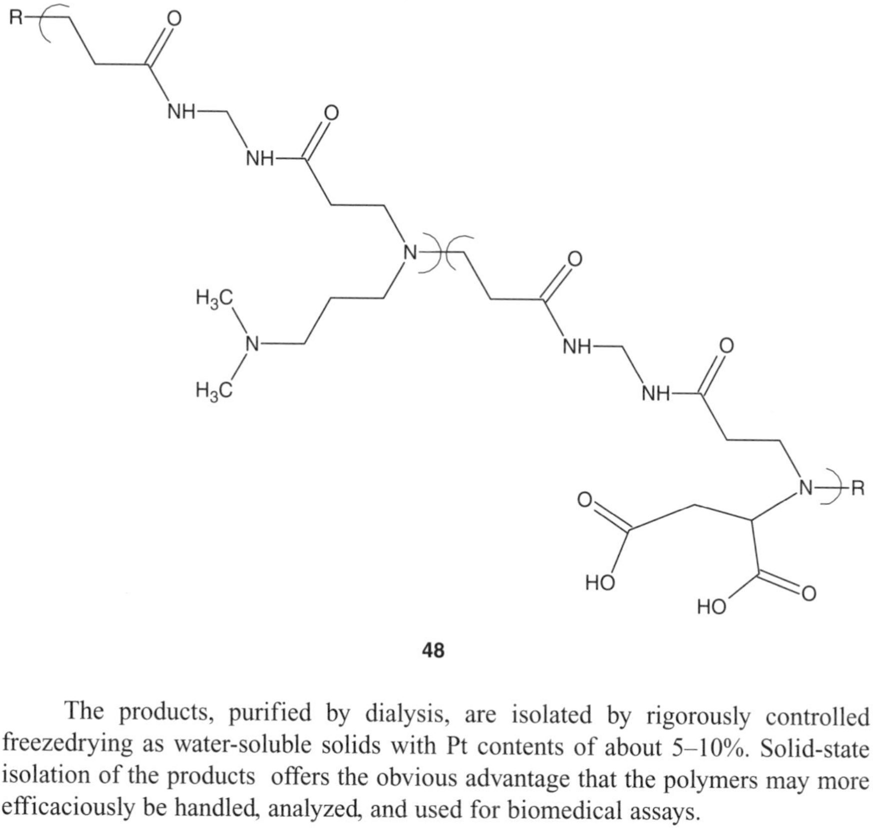

48

The products, purified by dialysis, are isolated by rigorously controlled freezedrying as water-soluble solids with Pt contents of about 5–10%. Solid-state isolation of the products offers the obvious advantage that the polymers may more efficaciously be handled, analyzed, and used for biomedical assays.

49

50

In a variant of these synthetic approaches, carriers bearing pairs of hydroxyl groups have been developed by Neuse and coworkers. A typical polymer containing the 1,2-dihydroxyl segment in the mainchain is shown in **51**. Again, the product will contain units that do not contain the DACH moiety.

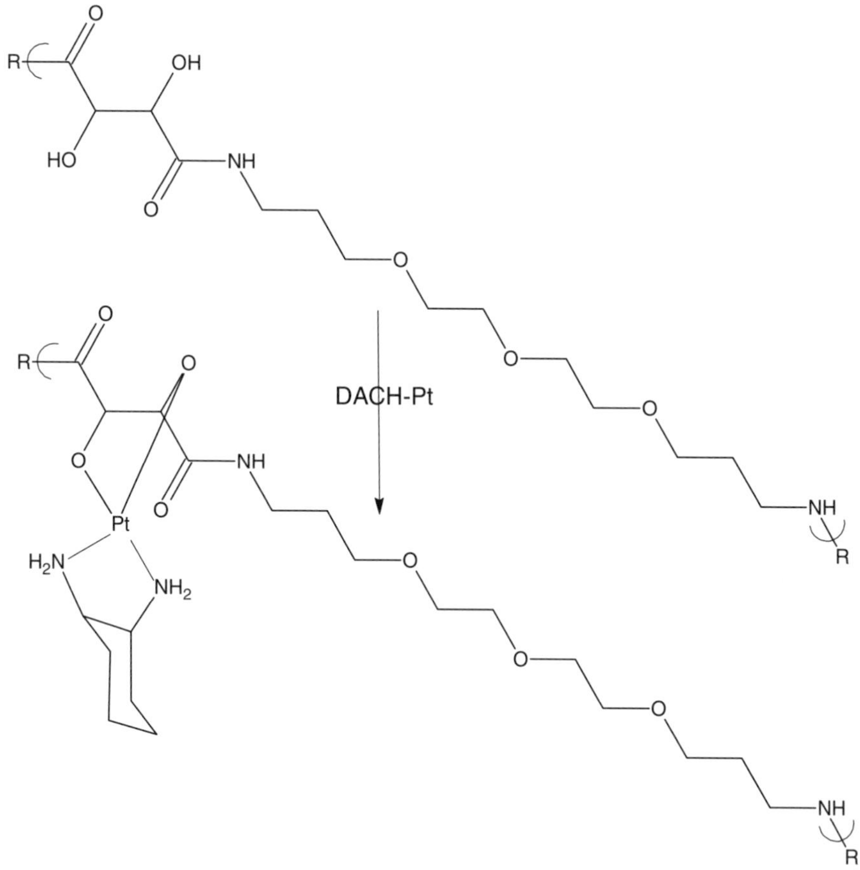

51

Another product where 1,2-diol functions are contained in sidechains is given in **52**.

Finally, a polymer featuring a 1-carboxyl-1-hydroxyl drug anchoring segment, which was synthesized by a two-step aminolysis of polysuccinimide, is given in **53**.

The results of a preliminary study of the in vitro performance of selected conjugates against the HeLa cell line are instructive.[142] Thus, two malonic acid derivatives of the general form **47** give IC_{50} values of 3.1 and 2.0 µg Pt/mL, quite respectable and comparable to results reported by Kataoka, Duncan, and others. Yet even better performances (IC_{50} = 0.60 µg Pt/mL) were found for the aspartic acid- modified polymer described as **49**. This polymer gave the same level of activity (IC_{50} = 0.43 mg Pt/mL) as found for product **50**. Of interest is the very similar behavior (IC_{50} = 0.51 µg Pt/mL)

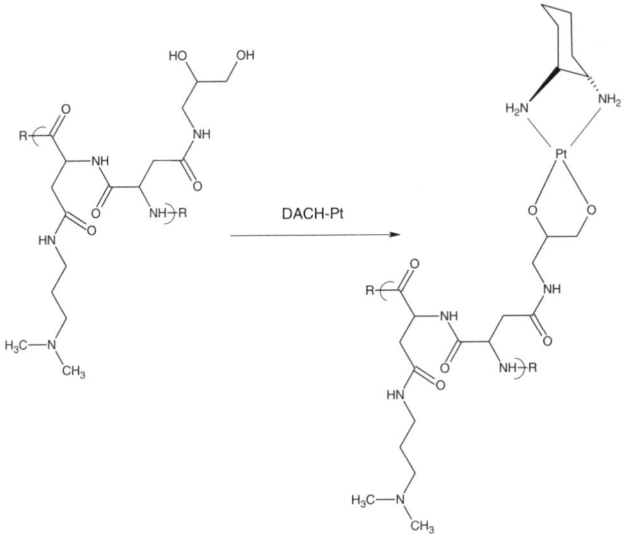

52

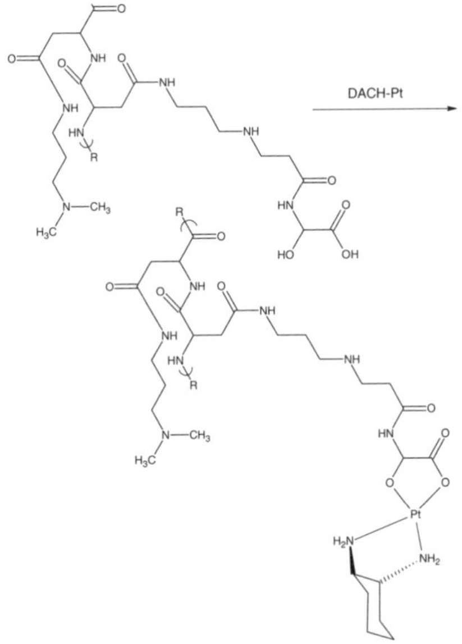

53

of the similar conjugate (not shown) featuring the (nonbasic) 2-hydroxyethyl solubilizing side group in place of the (strongly basic) 3-dimethylamino)propyl group, permitting the inference that in this type of conjugate the effect of the hydrosolubilizing entity

is not predominant. Finally, IC_{50} data for the conjugates with hydroxylatoplatinum coordination (structures **51–53**) have values of 0.26, 0.86, and 0.28 µg Pt/mL, respectively. Again, these excellent cytotoxic activities are on par with that of cisplatin itself (0.51 µg Pt/mL) assayed in the same test series for comparison. On balance, while these preliminary findings cannot be accepted as a basis for identifying detailed structure–performance relationships, they permit the conclusion that, at least in vitro, platinum anchoring via chelating leaving-group ligands provides a distinct therapeutic advantage. It remains to be seen whether in vivo screening will confirm these outstanding results.

A brief discussion of results appearing in patents follows since these are seldom cited in reviews. Patents discussed in this review illustrate the ongoing industrial interest in these compounds.

Gill and Andrulis[212] described the use of platinum amines complexed with water-soluble biodegradable or biostable polymers. The platinum is bonded covalently through the oxygen of the salt of the acid functionality as shown in **54**, producing the square–planar platinum

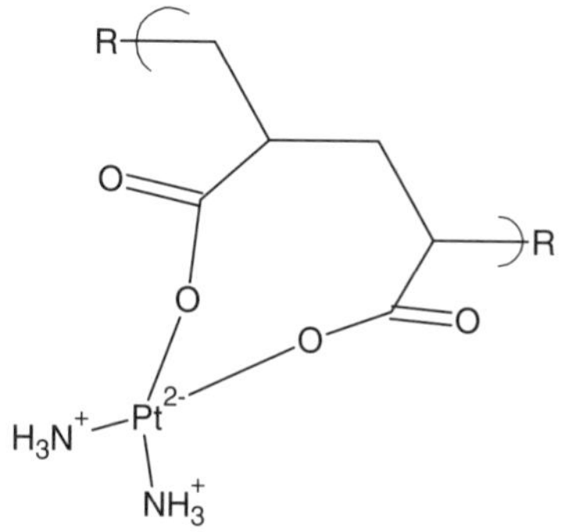

54

or octahedral platinum, **55**, depending on the reactant employed.

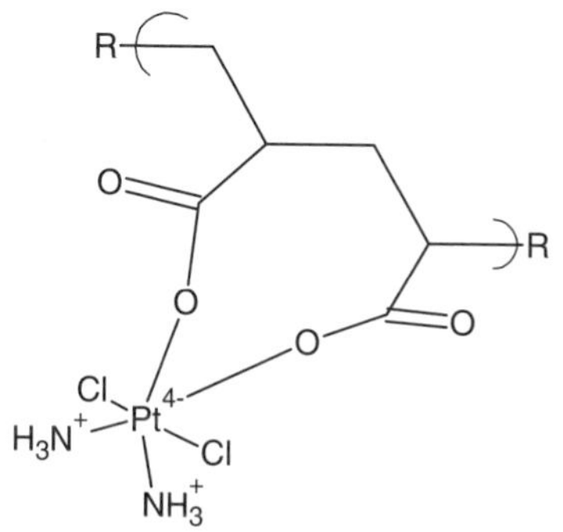

55

While the structures given above are for a simple ammonia derivative, the nitrogen-containing moiety on the platinum can be any monoamine. Also, for the octahedral complexes, while the chloride ligand is shown, this site can be filled by any anion. Suggested anions include sulfato, phosphato, halides, and pseudohalides. Structures of some of the synthetic polymers used are given in **56–61**.

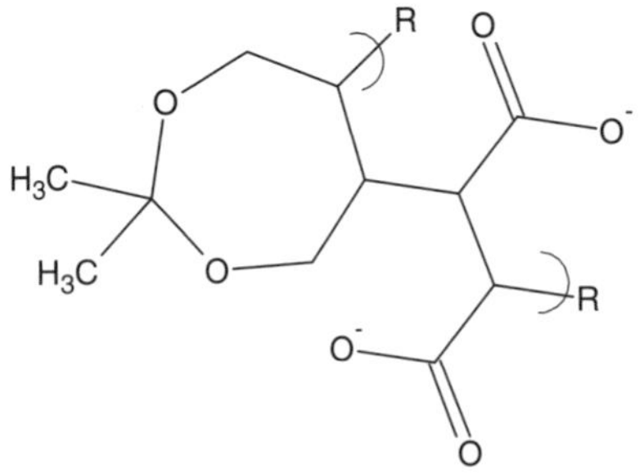

56

[poly(maleic anhydride-*co*-1,3-dimethyldioxepin), MA-DD]

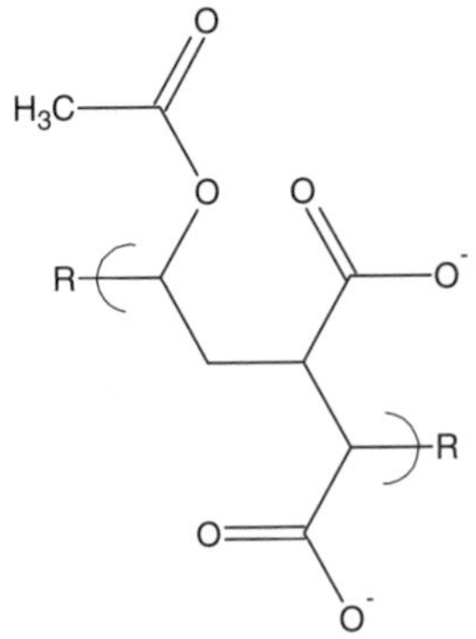

57

[poly(maleic anhydride-*co*-vinyl acetate), MA-VA]

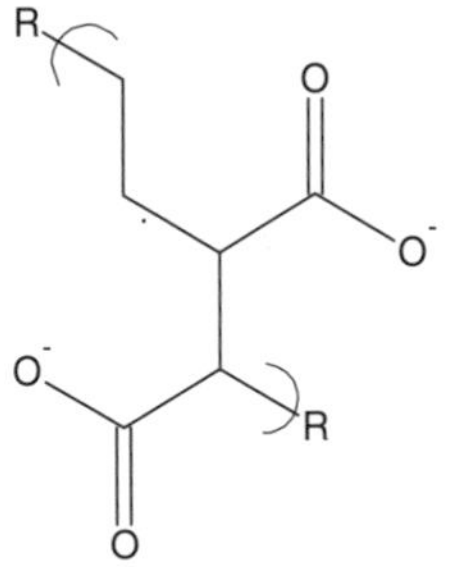

58

[poly(maleic anhydride-*co*-ethylene), MA-E]

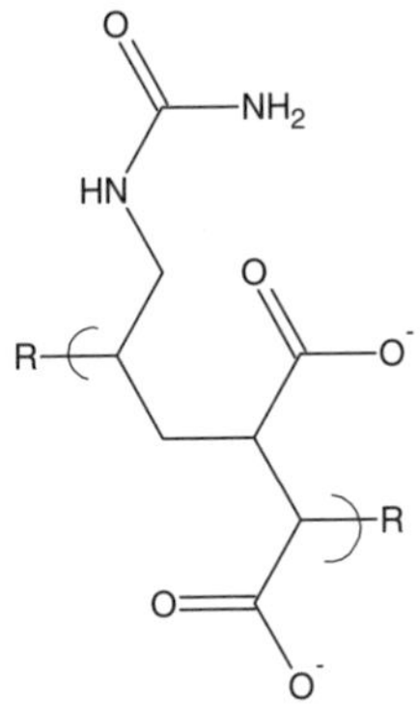

59

[poly(maleic anhydride-*co*-allyl urea), MA-UA]

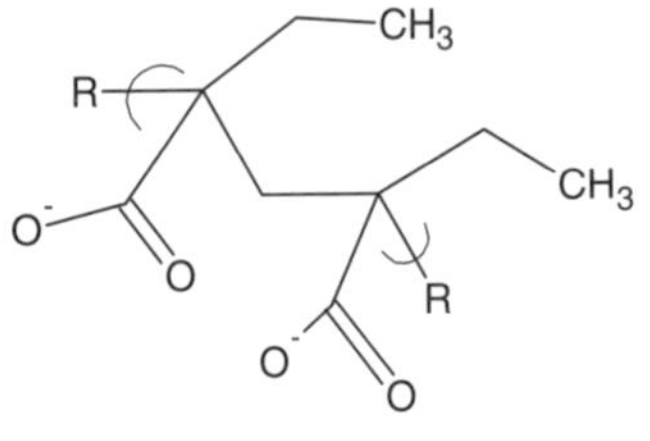

60

[poly(ethacrylic acid), EAA]

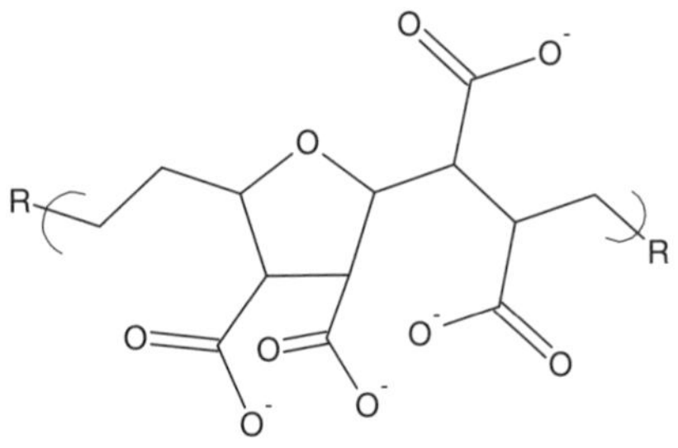

61

[poly(maleic anhydride-*co*-divinyl-ether), PYRAN]

Other possible components are a number of natural or naturally derived polymers such as carboxyamylose, carboxyamylopectin, carboxymannan, carboxypullulan, carboxydextrans, carboxydextran sulfate, carboxycellulose, oxidized carboxymethyl-cellulose, and carboxyalginic acid.

The platinum-containing reactants most used in this study[112] were *trans*-diaminocyclohexane salts, such as nitrates. These compounds are described as Pt(*trans*-DACH) complexes (**62**).

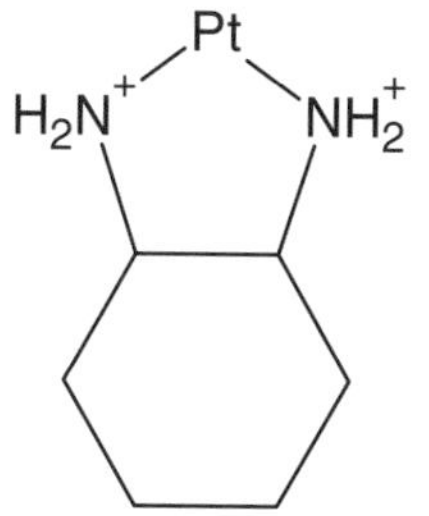

62

[*trans*-diaminocyclohexane salt]

At least some of the products are reported to be water-soluble to an extent equal to or greater than 1 mg/mL at a pH of 7 or higher, with some soluble to 700 mg/mL. Some tumor inhibition results were reported. For instance, the product from carboxyamylose gave 5/6 cures of BDF_1 mice infected with L1210 mouse leukemia at a dosage of 10 mg/kg, while the product from poly(acrylic acid) gave 1/6 cures in the same study. In contrast, cisplatin at 5 mg/kg showed no cures. In cell line tests, the product from Pt(*trans*-DACH) with carboxyamylose showed some to good inhibition of x5563 plasma cell myeloma, M5 ovarian carcinoma, Lewis lung, and CA-755 adenocarcinoma in the 5 mg/kg range.

Syntheses are generally carried out by dissolving both reactants in water at a pH of 4.6–6.5. Base, such as sodium hydroxide, is added if the system is too acidic. A 1.5–3-fold excess of the platinum reactant is usually employed. Because the platinum complex products are water soluble, the products were isolated using fractional precipitation or freeze drying. The excess uncomplexed polymer usually remains in solution if the reaction mixture is diluted with a water-soluble organic liquid such as ethanol or isopropanol until the polymer complex precipitates.

It is believed that the polymer complexes deliver greater doses of the platinum to the tumor site than is delivered using *cis*-DDP. It is not known if the mechanism of delivery of the platinum-containing moiety is the same or different from that of *cis*-DDP.

Drobnik and coworkers[213] worked on a similar theme. Various polymers including poly(amino acids), polysaccharides, and polyamides containing α-hydroxyacids with hydroxyl groups in the sidechains, were reacted with the acid chloride anhydride of trimellitic acid. This produces chains containing the following structure (**63**), given here for the product from poly(vinyl alcohol).

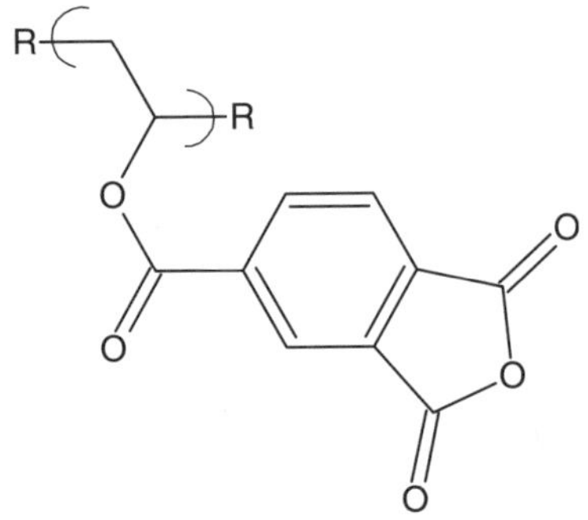

63

The anhydride group is then opened up presenting two acid groups as

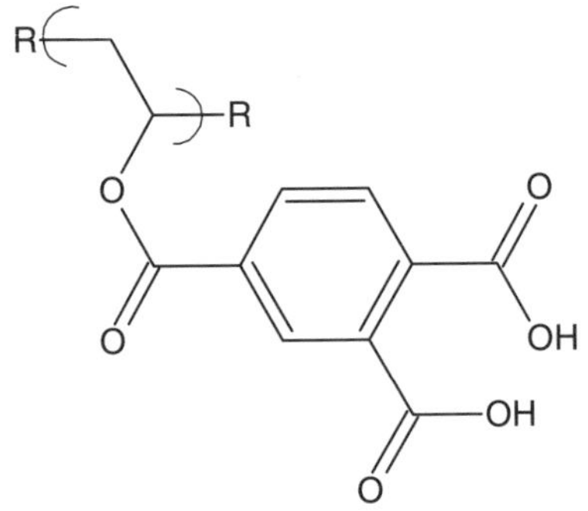

64

These acid groups are subsequently reacted with a platinum-containing moiety such as *trans*-1,2-diaminocyclohexane platinum(II) salt, forming the following platinum-containing polymer:

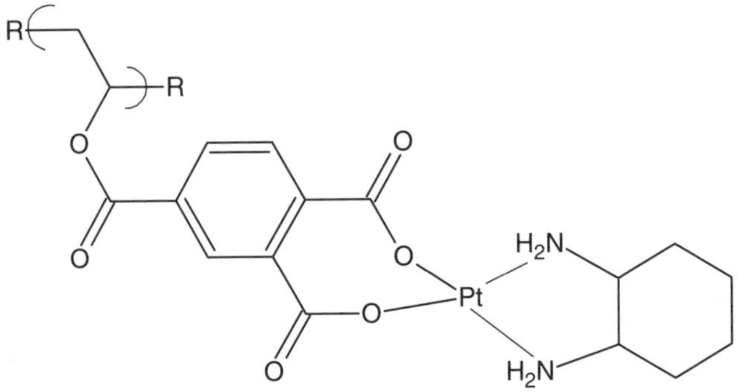

65

These reactions are similar to those described earlier, but they contain no additional solubilizing and targeting segments. In fact, this work probably acted as a basis for more elaborate tailoring with respect to the use of carboxylate reactive groups. Other amine–platinum reactants were also described.

Male SPF mice with leukemia P388 tumors were treated and evaluated for length of survival. The reference average survival time was 20 days. Treatment with dosages in the range of 10–62 μmol of platinum/kg resulted in survival times of 30–48 days for platinum-containing materials derived from poly[*N*-(2-hydroxymethyl)-D,L-aspartamide] and poly[*N*-(2-hydroxyethyl)-L-glutamine] both using *trans*-1,2-diaminocyclohexane platinum (II). Treatment of female mice H with ascitic tumors showed survival rates similar to and less than those found for *cis*-DDP itself.

A related study by Arnon, et al.[214] involved the formation of platinum complexes using polysaccharides and polyamino acids. Suggested carriers are dextran amine, carboxymethyl dextran, poly(L-glutamic acid), and poly(L-aspartic acid). The platinum compounds employed were potassium tetrachloroplatinate II, *cis*-DDP, and *cis*-Aq.

The *cis*-Aq (**66**) is obtained by reacting *cis*-DDP with silver nitrate.

$$\begin{array}{ccc} H_3N & & OH_2 \\ & \diagdown \; \diagup & \\ & Pt & \\ & \diagup \; \diagdown & \\ H_3N & & OH_2 \end{array}$$

cis-Aq

66

Reactions were generally carried out using aqueous solutions of the polymer carriers and three platinum-containing compounds, allowing the reaction to continue for 24 h. As expected, the amount of platinum bound varied with the ratios of reactants employed and factors such as reaction time. The particular type of platinum bonding was not well established and is probably a complex mixture of various units. It is highly likely that reaction with the amine-containing carriers resulted in formation of at least some Pt–N bonding but for those not containing amines, bonding presumably included formation of Pt–O bonds.

While *cis*-Aq is active against cancer cells, its toxicity is too high to allow its use in human therapy. Its in vivo toxicity appears to be due to its rapid binding capacity via exchange of water molecules. Potassium tetrachloroplatinum II is on the order of 20–100 times less active against cancer cells in vitro than *cis*-DDP and *cis*-Aq and it is essentially inactive against tumors in vivo. The authors, through binding these materials with the polymer carriers, were able to give products that showed moderate to good activity when tested in vivo against tumors including F9 embryonal carcinoma in Balb/c mice. The control mice died of the tumor at 34–67 days after tumor injection. Four of five tumor bearing mice died one day after receiving 100 μg of *cis*-DDP. Mice treated with three doses of 150 μg of *cis*-DDP complexed to the dextran amine did not die or develop tumors over the test period. Of five mice treated with 50 µg of *cis*-Aq, one died of drug toxicity, three died of the tumor and one survived. Four of the five mice treated twice with 150 µg of *cis*-Aq complexed to dextran amine survived the test protocol.

Iordanskaya et al.[215] reported results of *cis*-diamineplatinum complexed with a copolymer of sodium malenate and 1,4-diisopropoxy-2-butene that showed anticancer activity without nephrotoxicity or immunodepressant activity. These compounds appear similar in general structure to the compounds reported by Gill and Andrulis.[212]

XI. MIXED Pt–O/Pt–N-BOUND POLYMERS

A number of mixed products have been made. Some of these were previously reported in this chapter. Researchers emphasized either the Pt–N or Pt–O bonding, yet a mixture of bonding was probably present. Here we will focus on work emphasizing the mixture of Pt–N and Pt–O bonding performed by Rice and coworkers[216–219] of Access Pharmaceuticals. Much of this effort involves a DACH-platinum conjugate given the name AP along with a four-digit number.

AP 5280 contains *cis*-diamine platinum bound to a hydroxypropylmethacrylamide (HPMA) polymer via a tetrapeptide linker, comprising glycine–phenylalanine–leucine–glycine (GFLG) and an amidomalonic acid (ama) chelating group. The first product formed is the Pt–O product as seen in **67**. This is the kinetically preferred product.

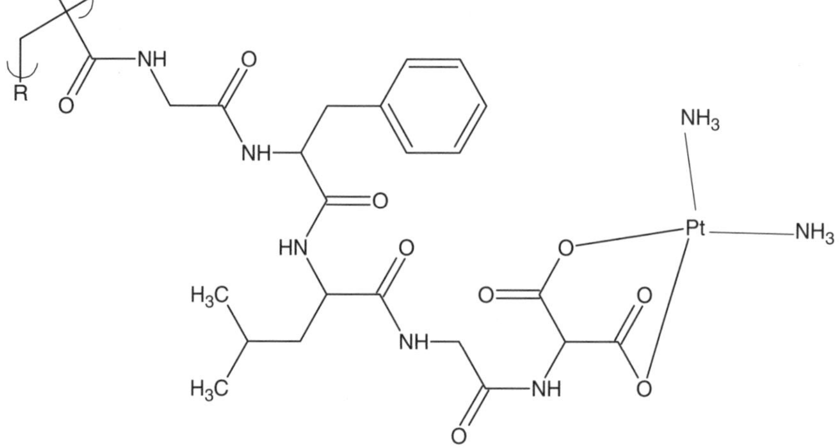

67

Heating converts the Pt–O to the Pt–N-bonded product **(68)**, the thermodynamically stable product.

AP 5286 is structurally similar to AP 5280 except that it has a *trans*-diaminocyclohexane, DACH, platinum chelate attached to the amine chelating group.

In comparison to carboplatin, the concentration of platinum in B16 melanoma tumors in mice is about 100-fold greater for AP 5280 for 20–120 h after IV administration of equivalent amounts of platinum. Pt tumor levels for AP 5280 are at least 10 times greater than those for carboplatin when dosed with equitoxic levels. Also, at equitoxic doses, AP 5280 generates a 10-fold greater number of B16 melanoma tumor Pt-DNA adducts compared to carboplatin.

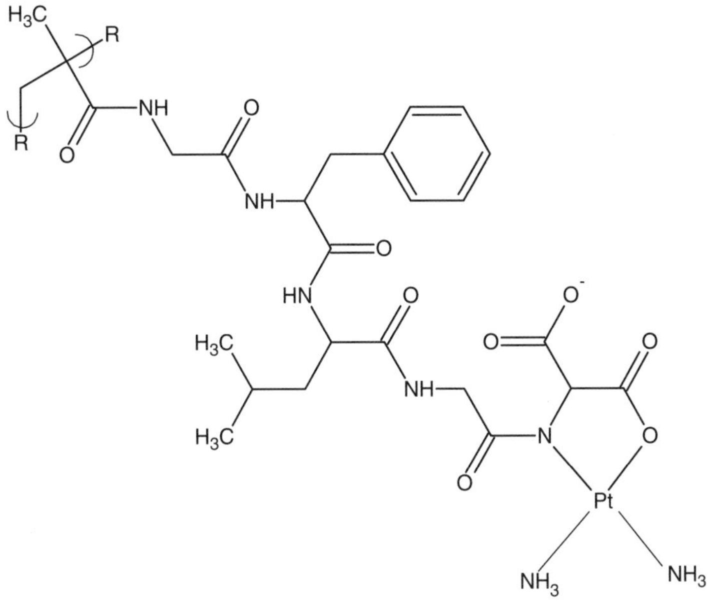

68

AP 5280 is also more effective than carboplatinum at inhibiting the growth of human tumor xenografts (UMSCC10b head and neck squamous cell tumors). At doses near its MTD, AP 5286 completely inhibits tumor growth for about 15 days. In a separate study, additional dosing allowed the complete inhibition of tumor growth to be extended. Similar results were found for AP 5286.

While structure **68** is cited, it is possible that the neutral structure **69** is also formed.

The Access Pharmaceuticals group has synthesized a number of other similar products that also show improved ability to inhibit various cancers, including Lewis lung carcinoma. One of these is AP 5346, which was made from poly(HPMA)-GG-ONp as described by Kopecek et al.[220] This polymer was coupled

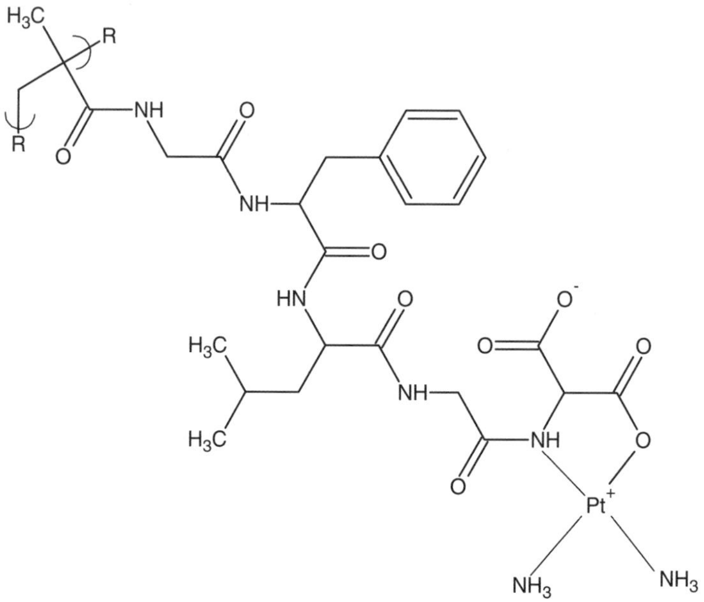

69

to diethyl *N*-glycylaminomalonate, and finally the DACH-PtCl$_2$ was used to plati-
nate the polymer. It was tested against oxaliplatin, the first DACH-platinum chelate
to enter routine clinical use. Oxaliplatin has a different spectrum of activity compared to cisplatin and carboplatin.[221] AP 6346 outperformed oxaliplatin in all of the
tests showing an improved efficacy and safety.

It is the intention of the group to develop both AP 5280 and AP 5346 because
the first one should have inhibition mechanisms similar to cisplatin and the second
to oxaliplatin.

XII. FUTURE WORK

It is clear that platinum-containing compounds, including polymeric materials,
are effective anticancer drugs. It is also clear that more needs to be done to define
the important factors in their activity. Mechanisms of cell inhibition need to be more
clearly defined. The mechanism of activity probably varies depending on the particular structure. Most likely, the compounds described in this chapter have several
different mechanisms of activity and such property–structure relationships need to
be established. Determining the actual mode (mechanism) and site of activity will
allow the development of a general structure–activity relationship. This may be the
greatest general need.

The use of "magic bullet" delivery systems should be an area of focus. There are a number of potential magic bullets that could be varied depending on the location of the cancer and site of desired anticancer activity. The use of such magic bullet delivery agents should limit negative side effects, thus reducing toxicity. Antibodies that preferentially bind tumor-associated or tumor-specific antigens are one possible direction. The antibody could encapsulate or be directly bonded to the platinum-containing moiety. Their use has been previously suggested and the general topic reviewed.[222, 223]

There are negative features to employing such antibodies: (1) the high variability of the internalization of antibody conjugates makes it dependent on both the cell and the antibody, (2) the antibody itself might be an antigen and stimulate an unwanted immune response in the patient, (3) the antibody might bind to normal cells that have the antigen recognized by the antibody, and (4) the attachment of platinum-containing groups onto the antibody may change its antigen-binding behavior.

Another avenue is the use of poly(amino acids) that have the platinum-containing moiety attached to them. This approach has already been used, but further study is warranted. Such poly(amino acids) might have several different drugs attached to them along with the platinum-containing moiety. The particular amino acid sequence might also be effective, for instance, at providing site direction to the drug, acting as a controlled-release agent, directing the cancer cell to be more available to drug treatment. The combination of the poly(amino acid) and antibody has also been tried with other drugs and might be one of the tools incorporated into the drug design.

Reedijk and coworkers[224] reported the use of solid-phase methods to produce cisplatin–amino acid products. This is described below with the final product given as **70**. Even though the products are not polymeric the approach illustrates how specially designed polypeptides can be connected to known platinum derivatives, allowing specific targeting to occur by utilizing known solid phase methods. They also reported the initial example of a solid-phase-mediated synthesis of peptide-tethered dichloroplatinium complexes[225] and the automated synthesis of a large family of analogs.[226]

Additional products synthesized by Reedijk are so-called dinuclear platinum complexes. In other work, such complexes have been found to offer high in vivo activity in both ciplatin and resistant tumors.[227,228]

The work of Reedijk and coworkers can be extended to the synthesis of mononuclear products as well as other polynuclear products.

Another delivery grouping includes bacterially derived products. Pasten et al.[229] suggested coupling Pseudomonas exotoxin with a variety of peptides, proteins, and growth factors that react with specific receptors on cells. Thus, they constructed a complex of *P. exotoxin* with a peptide hormone isolated from mice, epidermal growth factor, by introducing thiol groups into each and then linking the two using a disulfide exchange reaction.

Proteins have also been used as delivery agents. Thus, Meyers and Bichon[230] used a nonimmunoglobulin that preferentially binds to a tumor cell receptor in conjunction with a bonded biodegradable polymeric carrier that contains cytotoxic molecules.

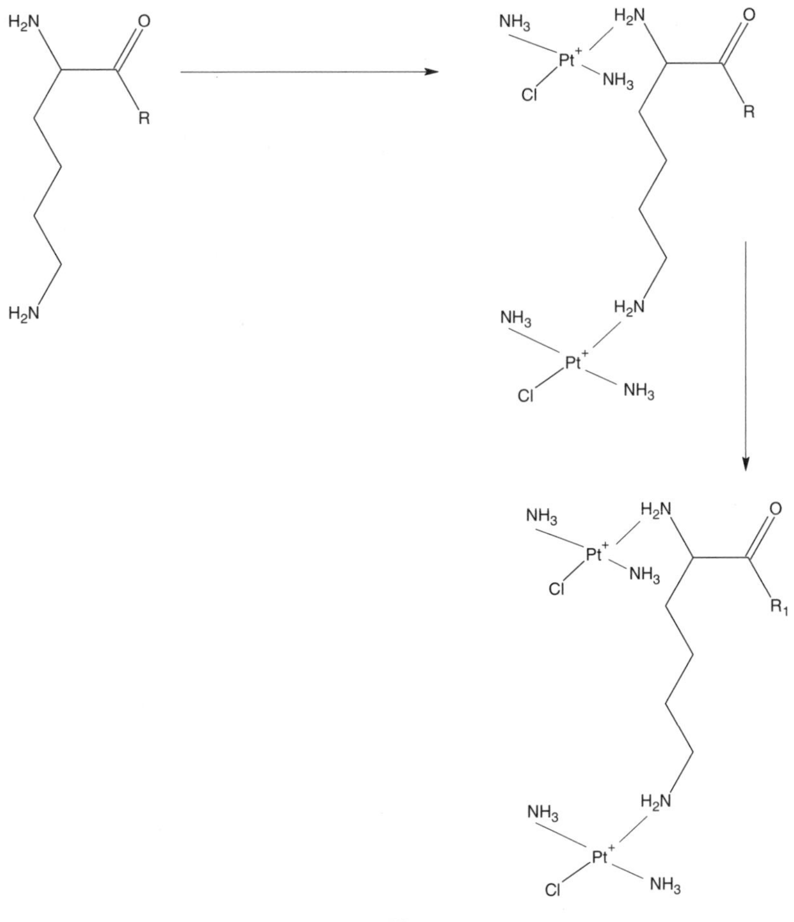

70

Thus, there are ample examples gleaned from related studies that should be applied to the appropriate delivery of platinum-containing drugs resulting in better drug delivery and more effective control of cancer.

XIII. ACKNOWLEDGMENTS

We are pleased to acknowledge the assistance of Professor Eberhard Neuse in preparing this chapter. He is instrumental in helping develop this area and, in fact, the entire area of metal-containing polymers. His pioneering work with ferrocene-containing polymers helped ignite the entire area of organometallic polymers and thus he is, in part, responsible for this series. We thank you.

XIV. REFERENCES

1. B. Rosenberg, L. Van Camp, T. Krigas, *Nature* (Lond.) **205**, 698 (1965).

2. M. Rozenwieg, D. VonHoff, M. Slavik, *Ann. Inst. Med.* **86**, 803 (1977).

3. I. Krakoff, *Cancer Treat. Rep.* **63**, 1523 (1979).

4. C. Litterset, I. Torres, A. Guarino, *J. Clin. Hem. Oncol.* **7**, 169 (1977).

5. G. Cadwell, E. Neuse, A. Perlwitz, *J. Inorg. and Organomet. Polym.* **7**, 111 (1997).

6. V. Murry, V., J. Whittaker, *J. Biochim. Biophys. Acta* **1354**, 261 (1997).

7. W. Hambley, *Coord. Chem. Rev.* **166**, 181 (1997).

8. L. Dalla Via, C. Noto, *Chem. Biol. Interact.* **110**, 203 (1998).

9. E. Neuse, *Polym. Adv. Technol.* **9**, 786 (1998).

10. G. Caldwell, E. Neuse, A. Perlwitz, *J. Appl. Polym. Sci.* **66**, 911 (1997).

11. C. Carraher, W. Scott, D. Giron, in *Bioactive Polymeric Systems*, C. Gebelein, C. Carraher, eds., Plenum, New York, 1985, Chpt. 20.

12. Y. Mika and M. Yokoyama, *Inorg. Chim. Acta* 51 (1998).

13. O. Heudi, A. Cailleus, *J. Inorg. Biochem.* **71**, 61 (1998).

14. D. Lebwohl, R. Canetta, *Eur. J. Cancer* **34**, 1522 (1998).

15. M. Tobe, J. Burgess, *Inorganic Reaction Mechanisms*, Longman, Essex, UK, 1999.

16. J. Huheey, E. Keiter, R. Keiter, *Inorganic Chemistry*, 4th ed., Haper Collins, New York, 1993.

17. J. Atwood, *Inorganic and Organometallic Reaction Mechanisms*, 2nd ed., VCH, New York, 1997.

18. F. Basolo, R. Pearson, *Prog. Inorg. Chem.* **4**, 381 (1964).

19. M. Korowski, D. Palmer, H. Kelm, *Inorg. Chem.* **18**, 2555 (1979).

20. U. Belluco, L. Cattalini, F. Basolo, R. Pearson, A. Turco, *J. Am. Chem. Soc.* **87**, 241 (1965).

21. F. Basolo, *Adv. Chem. Ser.* **49**, 81 (1965).

22. F. Basolo, R. Pearson, *Mechanisms of Inorganic Reactions*, Wiley, New York, 1968.

23. C. Langford, H. Gray, *Ligand Substitution Processes*, Benjamin, New York, 1965.

24. S. Hupp, G. Dahlgren, *Inorg. Chem.* **15**, 2349 (1976).

25. F. Basolo, J. Chatt, H. Gray, R. Pearson, B. Shaw, *J. Chem. Soc.* 2207 (1961).

26. R. Wilkins, *The Study of Kinetics and Mechanism of Reactions of Transition Metal Complexes*, Allyn & Bacon, Boston, 1974.

27. M. Tobe, *Inorganic Reaction Mechanisms*, Nelson, London, 1972.

28. R. Pearson, H. Sobel, J. Songstad, *J. Am. Chem. Soc.* **90**, 319 (1968).

29. U. Belluco, M.. Martelli, A. Orio, *Inorg. Chem.* **5**, 582 (1966).

30. R. Pearson, H. Gray, F. Basolo, *J. Am. Chem. Soc.* **82**, 787 (1960).

31. P. Haake, R. Pfeiffer, *Inorg. Chem.* **9**, 5243 (1970).

32. C. Carraher, G. Hess, W. Chen, *J. Polym. Mater.* **8**, 7 (1991).

33. C. Carraher, T. Manek, D. Giron, M. Trombley, K. Casberg, W. J. Scott, in *New Monomers and Polymers*, B. Culbertson, C. Pittman, eds., Plenum, New York, 1983.

34. C. Carraher, W. Chem, G. Hess, I. Lopez-Esbenshade, *Polym. Mater. Sci. Eng.* **78**, 104 (1998).

35. C. Carraher, W. Chem, G. Hess, D. Giron, *Polym. Mater. Sci. Eng.* **59**, 530 (1988).

36. C. Carraher, G. Hess, W. Chem, *Polym. Mater. Sci. Eng.* **59**, 744 (1988) and **58**, 557 (1988).

37. C. Carraher, W. Chem, G. Hess, D. Giron, in *Progress in Biomedical Polymers*, C. Gebelein, R. Dunn, eds., Plenum, New York, 1990.

38. C. Carraher, V. Nwufoh, J. R. Taylor, *Polym. Mater. Sci. Eng.* **60**, 685 (1989).

39. C. Carraher, A. Gasper, M. Trombley, F. Deroos, D. Giron, G. Hess, K. Casberg, in *New Monomers and Polymers*, B. Culbertson, C. Pittman, eds., Plenum, New York, 1983, Chapter 9.

40. J. Pesek, W. Mason, *Inorg. Chem.* **22**, 2958 (1983).

41. J. Rund, *Inorg. Chem.* **13**, 738 (1974).

42. R. Pearson, D. Johnson, *J. Am. Chem. Soc.* **86**, 3983 (1964).

43. D. Redfield, L. Cary, J. Nelson, *Inorg. Chem.* **14**, 50 (1975).

44. A. Verstuyft, L. Cary, J. Nelson, *Inorg. Chem.* **14**, 1501 (1976).

45. A. Versuyft, L. Cary, J. Nelson, *Inorg. Chem.* **15**, 3161 (1976).

46. D. Lebwohl, R. Canetta, *Eur. J. Cancer* **34**, 1522 (1998).

47. J. Dabrowiak, W. Bradner, *Progr. Med. Chem.* **24**, 129 (1987).

48. L. Kelland, *Crit. Rev. Oncol. Hematol.* **15**, 191 (1992).

49. M. Heim, *Metal Complexes in Cancer Chemotheraphy*, B. Keppler, ed., VCH, New York, 1992.

50. M. McKeage, L. Kelland, in *Molecular Aspects of Drug-DNA Interactions*, S. Neidle, M. Waring, eds., Macmillian, New York, 1992.

51. E. Neuse, *South Afr. J. Sci.* **95**, 509 (1999).

52. B. Rosenberg, *Cancer Treat. Rep.* **63**, 1433 (1979).

53. L. Zwelling, K. Kohn, *Cancer Treat. Rep.* **63**, 1439 (1979).

54. B. Rosenberg, *Platinum Metals Rev.* **15**, 42 (1971).

55. J. Roberts, J. Pascoe, *Advances in Antimicrobial and Antineoplastic Chemotherapy*, Vol. 2, Univ. Park Press, Baltimore, 1972, p. 249.

56. J. Drobnik, P. Horacek, *Chem. Biol. Interact.* **7**, 223 (1973).

57. A. Thomson, S. Mansy, *Advances in Antimicrobial and Antineoplastic Chemotherapy*, Vol. 2, Univ. Park Press, Baltimore, 1972, p. 199.

58. J. Macquet, T. Theophanides, *Inorg. Chim. Acta* **18**, 189 (1976).

59. S. Mansy, B. Rosenberg, A. Thomson, *J. Am. Chem. Soc.* **95**, 1633 (1973).

60. D. Goodgame, I. Jeeves, F. Phillips, *Biochim. Biophys Acta* **378**, 153 (1975).

61. J. Dehand, J. Jordanov, *J. Chem. Soc., Chem. Commun* 598 (1976).

62. A. Pegg, *Nature* **274**, 182 (1978).

63. D. Beck, J. Fisch, *Mut. Res.* **77**, 45 (1980).

64. L. Zwelling, K. Kohn, T. Anderson, *Proc. Am. Assoc. Res. ASCO* **19**, 233 (1978).

65. B. Rossenberg, *Cancer Treat. Rep.* **63**, 1433 (1979).

66. G. Cohen, W. Bauer, J. Barton, S. Lippard, *Science* **203**, 1014 (1979).

67. G. Cohen, J. Ledner, W. Baues, H. Ushay, C. Caravana, S. Lippard, *J. Am. Chem. Soc.* **102**, 2487 (1980).

68. J. Brouwer, P. van de Putter, A. Fichtinger-Schepman, J. Reedijk, *Proc. Natl. Acad. Sci. USA* **78**, 7010 (1981).

69. (a) B. Lippard, *Coord. Chem. Rev.* **182**, 263 (1999); (b) C. M. Sorenson, M. Berry, A. Eastman, *J. Natl. Cancer Inst.* **82**, 749 (1990); (c) G. Chu, *J. Bio. Chem.* **269**, 787 (1994).

70. M. Tobe, A. Khokhar, *J. Clin. Hematol. Oncol.* **7**, 114 (1977).

71. K. Matsumoto, M. Ochiai, *Coord. Chem. Rev.* **231**, 229 (2002).

72. J. Gottieb, B. Drewinko, *Cancer Chemother. Rep., Part 1* **59**, 621 (1975).

73. S. Stadnicki. R. Fleischman, U. Schaeppi, P. Merriman, *Cancer Chemother. Rep., Part 1* **59**, 467 (1975).

74. N. Farrell, L. R. Kelland, J. D. Roberts, M. van Beusichem, *Cancer Res.* **52**, 5065 (1992)

75. N. Farrell, Y. Qu, L. Feng, B. Vanhouten, *Biochemistry.* **29**, 9522 (1990).

76. N. Farrell, L. Kelland, J. Roberts, M. Beusichem, *Cancer Res.* **52**, 5065 (1992)

77. M. Coluccia, A. Boccarelli, M. A. Mariggio, N. Cardellicchio, P. Capreto, F. P. Intini, G. Natile, *Chem. Biol. Interact.* **98**, 251 (1995).

78. A. J. Kraker, J. D. Hoeschele, W. L. Elliot, H. D. Hollis-Showalter, A. D. Sercel, N. Farrell, *J. Med. Chem.* **35**, 4526 (1992).

79. (a) L. Kelland, *Drugs Future* **18**, 551 (1993); (b) K. Dorn, G. Hoerpel, H. Ringsdorf, in *Bioactuve Polymeric Systems*, C. Gebelein, C. Carraher, eds., Plenum, New York, 1985, Chapter 19; (c) H. Ringsdorf, *J. Polym. Sci. Polym. Symp.* **51**, 135 (1975); (d) C. J. T. Hoes, J. Feijen, in *Drug Carrier Systems*, F. Roerdink, A. Kroon, eds., Wiley, New York, 1989, p. 57; (e) H. Mlaeda, Y. Matsumura, *Crit. Rev. Ther. Drug Carrier Syst.* **6**, 193 (1989); (f) D. Putnam, J. Kopecek, *Adv. Polym. Sci.* **122**, 55 (1995); (g) R. Ducan, *Anti-Cancer Drugs* **3**, 175 (1992); (h) R. Mlaeda, *Adv. Drug Delivery Rev.* **6**, 181 (1991).

80. C. Carraher, in *Bioactive Systems*, C. Gebelein, C. Carraher, eds., Plenum, New York, 1985, Chapter 22.

81. D. Sigemann-Louda, C. Carraher, F. Pflueger, D. Nagy, J. Ross, in *Functional Condensation Polymers*, C. Carraher, G. Swift, eds., Kluwer, New York, 2002, Chapter 14.

82. C. Carraher, W. Scott, J. Schroeder, *J. Macromol. Sci. Chem.* **A15**(4), 625 (1981).

83. C. Carraher, D. Giron, I. Lopez, D. Cerutis, W. Scott, *Org. Coat. Plast. Chem.* **44**, 120 (1981).

84. D. Giron, M. Espy, C. Carraher, I. Lopez, in *Polymeric Materials in Medication*, C. Gebelein, C. Carraher, eds., Plenum, New York 1985, Chapter 14.

85. C. Carraher, R. Strothers, in *Applied Bioactive Polymeric Materials*, C. Gebelein, C. Carraher, V. Foster, eds., Plenum, New York, 1988.

86. D. Giron, M. Espy, C. Carraher, I. Lopez, C. Turner, *Polym. Mater. Sci. Eng.* **51**, 312 (1984).

87. C. Carraher, C. Ademu-John, D. Giron, J. Fortman, in *Metal-Containing Polymeric Systems*, J. Sheats, C. Carraher, C. Pittman, Plenum, New York, 1985, Chapter 11.

88. D. W. Sigemann, C. Carraher, A. Friend, *J. Polym. Mater.* **4**, 19 (1987).

89. D. W. Siegmann, C. Carraher, *J. Polym. Mater.* **4**,29 (1987).

90. D. W. Siegmann .D Brenner, C. Carraher, R. Strother, *Polym. Mater. Sci. Eng.* **61**, 214, 209 (1989).

91. D. W. Siegmann, D. Brenner, A. Colvin, B. Polner, R. Strother, C. Carraher, in *Inorganic and Metal-Containing Polymeric Materials*, J. Sheats, C. Carraher, C. Pittman, M. Zeldin, B. Currell, eds., Plenum, New York, 1990, p. 335.

92. D. W. Siegmann, D. Brenner, C. Carraher, *Polym. Mater. Sci. Eng.* **59**, 535 (1988).

93. D. W. Siegmann, C. Carraher, D. Brenner, in *Progress in Biomedical Polymers*, C. Gebelein, R. Dunn, eds., Plenum, New York, 1990, p. 371.

94. D. W. Siegmann, C. Carraher, A. Friend, *Polym. Mater. Sci. Eng.* **56**, 79 (1987).

95. D. W. Siegmann-Louda, C. Carraher, F. Pflueger, D. Nagy, J. Ross, in *Functional Condensation Polymers*, C. Carraher, G. Swift, eds., Kluwer, New York, 2002 Chapter 14.

96. C. Carraher, L. Tissinger, M. Williams, I. Lopez, *Polym. Mater. Sci. Eng.* **58**, 239 (1988).

97. C. Carraher, L. Tissinger, I. Lopez, M. Williams, in *Biomimetic Polymers*, C. Gebelein, ed., Plenum, New York, 1990.

98. C. Carraher, M. Williams, in *Cosmetic and Pharmaceutical Applications of Polymers*, C. Gebelein, T. Cheng, V. Yang, eds., Academic Press, New York, 1991.

99. L. M. Volshtein, I. Lukyanova, *Neorg. Khim.* **11**(6), 708 (1966)

100. L. M. Volshtein, L. Dikanskaya, *Russ. J. Inorg. Chem.* **13**(9), 1304 (1968); **19**(1), 81 (1974).

101. L. Grantham, T. Elleman, D. Martin, *J. Am. Chem. Soc.* **77**, 2966 (1955).

102. F. Hartley, *The Chemistry of Platinum and Palladium*, Applied Science, London, 1972.

103. C. Carraher, T. Manek, M. Trombley, G. Hess, D. Giron, *ACS Polym. Preprints* **24**(1) (1983).

104. C. Carraher, T. Manek, D. Gorin, D. Cerutis, M. Trombley, *ACS Polym. Preprints* **23**(2), 77 (1982).

105. C. Carraher, T. Manek, D. Giron, M. Trombley, K. Casberg, W. Scott, *New Monomers and Polymers*, B. Culbertson, C. Pittman, eds., Plenum, New York, 1983, Chapter 8.

106. C. Carraher, R. Strothers, D. Brenner, *Polym. Mater. Sci. Eng.* **57**, 173 (1987).

107. C. Carraher, R. Strothers, in *Applied Bioactive Polymeric Materials*, C. Gebelein, C. Carraher, V. Foster, eds., Plenum, New York, 1988, Chapter 11.

108. C. Carraher, I. Lopez, *Polym. Mater. Sci. Eng.* **54**, 618 (1986).

109. C. Carraher, N. Bigley, M. Trombley, D. Giron, *Polym. Mater. Sci. Eng.* **57**, 177 (1987).

110. C. Carraher, I. Lopez, D. Giron, *Advances in Biomedical Polymers*, C. Gebelein, ed., Plenum, New York, 1987.

111. C. Carraher, I. Lopez, D. Giron, *Polym. Mater. Sci. Eng.* **53**, 644 (9185).

112. D. Giron, M. Espy, C. Carraher, I. Lopez, in *Polymeric Materials in Medication*, C. Gebelein, C. Carraher, eds., Plenum, New York, 1985, Chapters 13, and 14.

113. H. R. Allcock, R. Allen, J. O'Brien, U.S. Patent 4,151,185 (April. 24, 1979); H. Allcock, R. Allen, J. O'Brien, *JCS Chem. Commun.*, 717 (1976).

114. S. Song, Y. Sohn, *Proc.* 24th *Int. Symp. Controlled Release Bioactive Materials*, 1997, p. 551.

115 C. Carraher, C. Ademu-John, J. Fortman, D. Giron, R. Linville, *Polym. Mater. Sci. Eng.* **49**, 210 (1983).

116. C. Carraher, C. Ademu-John, J. Fortman, D. Giron, C. Turner, *J. Polym. Mater.* **1**, 116 (1984).

117. C. Carraher, C. Ademu-John, J. Fortman, D. Giron, in *Metal-Containing Polymeric Systems*, J. Sheats, C. Carraher, C. Pittman, Eds., Plenum, New York, 1985.

118. C. Carraher, C. Ademu-John, J. Fortman, D. Giron, C. Turner, R. Linville, in *Polymeric Materials in Medication*, C. Gebelein, C. Carraher, eds., Plenum, New York, 1985.

119. C. Carraher, C. Ademu-John, J. Fortman, D. Giron, C. Turner, *Polym. Mater. Sci. Eng.* **51**, 307 (1984)

120. C. Carraher, A. Francis, D. Siegmann-Louda, in *Functional Condensation Polymers*, C. Carraher, G. Swift, eds., Kluwer, New York, 2002, Chapter 15.

121. C. Carraher, A. Francis, D. Siegmann-Louda, *Polym. Mater. Sci. Eng.* **84**, 654, 664 (2001).

122. B. Howell, E. Walles, R. Rashidianfar, *Makromol. Chem., Macromol. Symp.* **19**, 329 (1988).

123. B. Howell, E. Walles, *Inorg. Chim. Acta* **142**, 185 (1988).

124. B. Howell, E. Walles, *ACS Polym. Preprints* **27**, 460 (1986).

125. B. Howell, E. Walles, U.S. Patent 4,405,757 (1983).

126. R. Arnon, E. Hurwitz, R. Kashi, M. Wilchek, M. Sela, B. Schechter, Eur. Patent. 99133 (Nov. 1988); Appl. EP 83-106965 (1983).

127. W. Shen, K. Beloussow, M. Meirim, E. Neuse, G. Caldwell, *J. Inorg. Organomet. Polym.* **10**, 51 (2000).

128. B. Schechter, G. Caldwell, M. Meirim, E. Neuse, *Appl. Organomet. Chem.* **14**, 701 (2000).

129. G. Caldwell, E.W. Neuse, C. VanRensburg, *Applied Organomet. Chem.* **13**, 189 (1999).

130. E. W. Neuse, *Polym. Adv. Technol.* **9**, 786 (1998).

131. G. Caldwell, E. Neuse, C. VanRensburg, *J. Inorg. Organomet. Polym.* **7**, 217 (1997).

132. E. W. Neuse, G. Caldwell, *J. Inorg. Organomet. Polym.* **7**, 163 (1997)

133. E. W. Neuse, G. Caldwell, A. Perlwitz, *Polym. Adv. Technol.* **7**, 867 (1996).

134. E. W. Neuse, G. Caldwell, A. Perlwitz, *J. Inorg Organomet. Polym.* **5**, 195 (1995).

135. E. W. Neuse, *Macromol. Symp.* 80, 111 (1994).

136. U. Chiba, E. Neuse, J. Swarts, G. Lamprecht, *Angew. Makromol. Chem.* **214**, 137 (1994).

137. C. Mbonyana, E. W. Neuse, A. Perlwitz, *Appl. Organomet. Chem.* **7**, 279 (1993).

138. E. W. Neuse, B. Patel, C. Mbonyana, W. Carol, *J. Inorg, Organomet. Polym.* **1**, 147 (1991).

139. G. Caldwell, E. W. Neuse, A. Stephanou, *J. Appl. Polym. Sci.* **50**, 393 (1993).

140. M. de L. Machado, E. W. Neuse, A. Perlwitz, S. Schmitt, *Polym. Adv. Technol.* **1**, 275 (1990).

141. G. Caldwell, E. W. Neuse, A. Perlwitz., *J. Appl. Polym. Sci.* **54**, 57 (1994).

142. E. W. Neuse, C. E. J. van Rensburg, unpublished results.

143. J. Drobnik, V. Saudek., J. Flasak, J. Kalal, *J. Polym. Sci., Polym. Symp.* **66**, 65 (1979).

144. P. Neri, G. Antoni, *Macromol. Synth.*, **8**, 25 (1982).

145. E. Gianasi, M. Wasil, E. Evagorou, A. Keddle, G. Wilson, R. Duncan, *Eur. J. Cancer* **35**, 994 (1999).

146. R. Duncan, E. Evagorou, R. Buckley, E. Gianasi, M. Wasil, G. Wilson, *Proc. Int. Symp. Controlled Release Bioactive Materials*, 24th, 1997, p. 775.

147. R. Duncan, M. Wasil, E. Gianasi, A. Keddle, E. Evagorous, R. Buckley, G. Wilson, *Proc. 24th Int. Symp. Controlled Release Bioactive materials*, 1997, p. 83.

148. Y. Kidani, M. Noji, T. Tashiro, *Gann* **71**, 637 (1980).

149. Y. Kidani, K. Inagaki, R. Saito, S. Tsukagoshi, *J. Clin. Hematol. Oncol.* **7**, 197 (1977).

150. Y. Chao, A. Holtzner, S. Mastin, *J. Am. Chem. Soc.* **99**, 8024 (1977).

151. R. Arnon, B. Schechter, M. Wilchek, *Adv. Exp. Med. Biol.* **303**, 79 (1991).

152. B. Schechter, R. Arnon, M. Wilchek, J. Schlessinger, E. Aboud-Pirak, M. Sela, *Int. J. Cancer* **48**, 167 (1991).

153. B. Schechter, A. Neumann, M. Wilchek, R. Arnon, *J. Controlled Release* **10**, 75 (1989).

154. B. Schechter, M. Rosing, M. Wilchek, R. Arnon, *Cancer Chemother. Pharmacol.* **24**, 161 (1989).

155. B. Schechter, R. Pauzner, R. Arnon, J. Haimovich, M. Wilchek, *Cancer Immunol. Immunother.* **25**, 225 (1987).

156. B. Schechter, M. Wilchek, R. Arnon, *Int. J. Cancer* **39**, 409 (1987).

157. B. Schechter, R. Pauzner, M. Wilchek, R. Arnon, *Cancer Biochem. Biophys.* **8**, 289 (1986).

158. R. Arnon, B. Schechter, M. Wilchek, *Adv. Exp. Med. Biol.* **303**, 79 (1991).

159. B. Schechter, R. Arnon, M. Wilchek, J. Schlessinger, E. Hurwitz, E. Aboud-Pirak, M. Sela, *Int. J. Cancer* **48**, 167 (1991).

160. B. Schechter, M. Rosing, M. Wilchek, R. Arnon, *Cancer Chemoth. Pharmacol.* **24**, 161 (1989).

161. B. Schechter, R. Pauzner, R. Arnon, M. Wilchek, *Cancer Biochem. Biophys.* **8**, 277 (1986).

162. D. Avichezer, B. Schechter, R. Arnon, *React. Funct. Polym.* **36**, 59 (1998).

163. L. Chen, B. Schechter, R. Armon, M. Wilchek, *Drug Devel. Res.* **50**, 258 (2000).

164. A. Breskoin, R. Chechik, Z. Paltiel, B. Schechter, A. Warshawsky, A. Shanzer, M. Neeman, EU Patent 991430 (1998).

165. B. Schechter, R. Arnon, M. Wilchek, *React. Polym.* **25**, 167 (1995).

166. B. Schechter, R. Arnon, Y. Freedman, L. Chen, M. Wilchek, *J. Drug Targ.* **4**, 171 (1996).

167. K. Fujii, T. Imai, S. Shiraishi, M. Otagiri, *Pharm. Sci.* **2**, 475 (1996).

168. T. Imai, K. Fujii, S. Shiraishi, M. Otagiri, *J. Pharm. Sci.* **86**, 244 (1997).

169. K. Fujii, T. Imai, M. Otagiri, *Proc. Int. Symp. Controlled Release of Bioactive Materials*, (1996), p. 639.

170. T. Imai, K. Fujii, S. Shiraishi, M. Otagiri, *J. Pharm. Sci.* **86**, 244 (1997).

171. N. Malik, R. Duncan, WO Patent 98/47537 (Oct. 1998).

172. R. Duncan, N. Malik, E. Evagorou, R. Duncan, *Anti-Cancer Drugs* **10**, 767 (1999).

173. P. Ferruti, E. Ranucci, F. Trotta, E. Gianasi, E. Evagorou, M. Wasil, G. Wilson, R. Ducan, *Macromol. Chem. Phys.* **200**, 1644 (1999).

174. R. Duncan, E. Evagorou, R. Buckley, E. Gianasi, U.S. Patent Appl. 1998-60455 (1998).

175. R. Duncan, E. Evagorou, R. Buckley, E. Gianasi, WO Patent 1998-US6770 (1998).

176. R. Duncan, P. Ferruti, E. Evagorou, WO Patent 1998-US7659 (1998).

177. N. Malik, E. Evagorou, R. Ducan, *Proc. Int. Symp. Controlled Release Bioactive Materials*, 1997, p. 107.

178. R. Duncan, E. Evagorou, R. Buckley, E. Gianasi, Jpn. Appl. Patent JP 1998-546083 (1998).

179. Y. Ohya, H. Oue, K. Nagatomi, T. Ouchi, *Biomacromolecules* **2**, 927 (2001).

180. I. Katsuro, N. Tomiyama, M. Nakashima, Y. Ohya, M. Ichikawa, T. Ouchi, T. Kanematsu, *Anti-Cancer Drugs* **11**, 33 (2000).

181. M. Nakashima, K. Ichinose, T. Kanematsu, T. Masunaga, Y. Ouchi, N. Tomiyama, H. Sasaki, M. Ichikawa, *Biol. Pharm. Bull.* **22**, 756 (1999).

182. Y. Ohya, T. Masunaga, T. Ouchi, K. Ichinose, M. Nakashima, M. Ichikawa, T. Kanematsu, *Tailored Polymeric Materials for Controlled Delivery Systems*, ACS, Washington, DC, p. 266.

183. Y. Ohya, T. Masunaga, T. Ouchi, K. Ichinose, M. Nakashima, M. Ichikawa, T. Kanematsu, *ACS Polym Preprints* **38**, 606 (1997).

184. Y. Ohya, T. Masunaga, T. Ouchi, K. Ichinose, M. Nakashima, M. Ichikawa, T. Kanematsu, *Proc 24th Int. Symp. Controlled Release of Bioactive Materials*, 1997, p. 755.

185. Y. Ohya, T. Masunaga, T. Baba, T. Ouchi, *J. Biomater. Sci., Polym. Ed.* **7**, 1085 (1996).

186. Y. Ohya, T. Masunaga, T. Baba, T. Ouchi, *J. Macromol. Sci., Pure Appl. Chem.* **A33**, 1005 (1996).

187. K. Ichinose, N. Tomiyama, M. Nakashima, Y. Ohya, M. Ichikawa, T. Ouchi, T. Kanematsu, *Anti-Cancer Drugs* **11**, 33 (2000).

188. M. Nakashima, K. Ichinose, T. Kanematsu, T. Masunaga, Y. Ohya, T. Ouchi, N. Tomiyama, H. Sasaki, M. Ichikawa, *Biol. Pharm. Bull.* **22**, 756 (1999).

189. Y. Ohya, T. Masunaga, T. Baba, T. Ouchi, *J. Biomater. Sci., Polym. Ed.* **7**, 1085 (1996).

190. Y. Ohya, S. Shirakawa, M. Matsumoto, T. Ouchi, *Polym. Adv. Technol.* **11**, 635 (2000).

191. M. Yokoyama, T. Okano, Y. Sakurai, S. Suwa, K. Kataoka, *J. Controlled Release* **39**, 351 (1996).

192. K. Kataoka, N. Nishiyama, M. Yokoyama, T. Okano, Jpn. Patent WO 2001-JP8337 (2001).

193. G. Kwon, M. Yokoyama, T. Okano, T. Sakurai, K. Kataoka, *Pharm. Res.* **10**, 970 (1993).

194. N. Nishiyama, Y. Kato, Y. Sugiyama, K. Kataoka, *Pharm. Res.* **18**, 1035 (2001).

195. N. Kas'yanenko, E. Aia, A. Bogdanov, Y. Kosmotynskaya, K. Yakoviev, *Mol. Biol.* **36**, 594 (2002).

196. N. Kas'yanenko, A. Bogdanov, S. Defrenne, *Biofizika* **47**, 449 (2002).

197. A. Bogdanov, C. Martin, A. Bogdanova, T. Brady, R. Weissleder, *Bioconj. Chem.* **7**, 144 (1996).

198. A. Bogdanov, R. Weisslender, T. Brady, U.S. Pat. 5,8717,710 (1996).

199. A. Bogdanov, S. Wright, E. Marecos, A. Bogdanova, C. Martin, P. Petherick, R. Weissleder, *J. Drug Targ.* **4**, 321 (1997).

200. M. J. Han, S. Park, T. Cho, J. Chang, *Polymer* **37**, 667 (1996).

201. M. J. Han, N. Cho, T. Cho, J. Chang, *J. Polym. Sci., Part A, Polym. Chem.* **33**, 1829 (1995).

202. M. J. Han, W. Choong, K. Kim, S. Lee, *Macromolecules* **25**, 3528 (1992).

203. M. J. Han, B. Kae, H. Ki, J. Tae, Y. Won, W. Shim, *J. Bioact. Compat. Polym.*, **5**, 420 (1990).

204. M. J. Han, K. Kim, T. Cho, K. Choi, *J. Polym. Sci., Part A, Polym. Chem.* **28**, 2719 (1990).

205. M. J. Han, K. Choi, J. Chae, B. Hahn, W. Lee, *J. Bioact. Compat. Polym.* **5**, 90 (1990).

206. M. J. Han, S. Kang, W. Lee, *Bull. Kor. Chem. Soc.* **11**, 154 (1990).

207. M. J. Han, D. Lee, W. Lee, B. Hahn, *Bull. Kor. Chem. Soc.* **10**, 212 (1989).

208. M. J. Han, T. Cho, S. Park, Y. Sown, C. Lee, S. Choi, *J. Bioact. Compat. Polym.* **7**, 358 (1992).

209. M. Han, S. Park, T. Cho, J. Chang, Y. Sohn, C. Lee, *J. Bioact. Compat. Polym.* **9**, 142 (1994).

210. Z. Wang, Chinese *J. Polym. Sci.* **7**, 159 (1989).

211. E. W. Neuse, N. Mphephu, H. Netshifhefhe, M. Johnson, *Polym. Adv. Technol.* (in press).

212. D. Gill, P. J. Andrulis, U.S. Patent 4,931,553 (June 5, 1990).

213. J. Drobnik, D. Noskova, F. Rypacek, M. Metalova, V. Saudek, U.S. Patent 4,659,849 (April 21, 1987).

214. R. Arnon, M. Wilchek, B. Schechter, Eur. Patent 190,464 (1985).

215. L. Iordanskaya, L. Stotskaya, Y. Ulogova, B. Krentsel, L. Gubajdulin, Y. Guryshev, I. Starshinova, I. Rusakov, K. Shchitkov, G. Sukhin, E. Percherskikh, V. Chissov, A. Kozlov, L. Reznik, Y. Natochin, N. Rastova, V. Yurev, Russ. Patent 2034856 (May 10, 1995).

216. J. Rice, D. Stewart, R. Safaei, S. Howell, D. Nowotnik, paper presented at 5th Int. Symp. Polym. Therapeutics, Wales, Jan. 3–5, 2002.

217. D. Stewart, J. Rice, P. Sood, J. St. John, S. Thurmond, D. Nguyen, G. Russell-Jones, D. Nowotnik, paper presented at 2nd Int. Symp. Tumor Targeted Delivery Systems, Rockville, MD, Sept. 22–25, 2002.

218. M. Tibben, J. Rademaker-Lakhai, J. Rice, D. Stewart, J. Schellens, J. Beijnen, *Anal. Bioanal. Chem.* **373**, 233 (2002).

219. M. Bouma, B. Nuijen, D. Stewart, J. Rice, B. Jansen, J. Reedijk, A. Bult, J. Beijnen, *Anti-Cancer Drugs* **13**, 915 (2002).

220. J. Kopecek et al., U.S. Patent 5,037,883.

221. L. R. Kelland, Crit. *Rev. Oncol. Hematol.* **15**, 191 (1993).

222. K. Sikora, *Br. Med. Bull.* **40**, 233 (1984).

223. A. Myers, D. Bichon, U.S. Patent 5,087,616 (Feb. 11, 1992).

224. S. van Zutphen, M. S. Robillard, G. van der Marel, H. S. Overkleeft, H. den Dulk, J. Brouwer, J. Reedijk, *Chem. Commun.* 634 (2003).

225. M. S. Robillare, A. Valentijy, N. Meeuwenoord, G. van der Marel, J. van Boom, J. Reedijk, *Angew. Chem., Int. Ed. Engl.* **39**, 3069 (2000).

226. M. Robillare, M. bacac, H. van der Elst, A. Flamigni, G. van der Marel, J. van Boom, J. Reedijk, *J. Comb. Chem.* (in press).

227. A. Kraker, J. Hoeschele, W. Elliott, H. Showalter, A. Sercel, NB. Farrell, *J. Med. Chem.* **35**, 4526 (1992).

228. Y. Qu, N. Farrell, *J. Am. Chem. Soc.*, **113**, 4851 (1991).

229. I. Pasten, M. Willingham, D. Fitzgerald, U.S. Patent 4,545,985 (1986).

230. A. Myers, D. Bichon, U.S. Patent 5,087,616 (Feb. 11, 1992).

CHAPTER 8

New Organic Polyacid–Inorganic Composites for Improved Dental Materials

Bill M. Culbertson, Minhhoa H. Dotrong, and Scott R. Schricker

College of Dentistry, The Ohio State University, Columbus, Ohio

CONTENTS

I. INTRODUCTION   194

II. GLASS IONOMER TECHNOLOGY   194
    A. Amino Acid–Modified Glass Ionomers   198
    B. *N*-Vinylpyrrolidone (NVP)-Modified Glass Ionomers   201

III. NEW NVP-MODIFIED GLASS IONOMERS: EXPERIMENTAL WORK   202
    A. Materials   202
    B. Polymer Synthesis   203
    C. Characterization   203
    D. Physical Properties   203

IV. RESULTS AND DISCUSSION   204

V. CONCLUSIONS   205

VI. REFERENCES   206

Macromolecules Containing Metal and Metal-Like Elements,
Volume 3: Biomedical Applications, edited by Alaa S. Abd-El-Aziz,
Charles E. Carraher Jr., Charles U. Pittman Jr., John E. Sheats, and Martel Zeldin
ISBN: 0-471-66737-4   Copyright © 2004 John Wiley & Sons, Inc.

I. INTRODUCTION

Water-soluble polymers having pendant carboxylic acid groups, such as a modified poly(acrylic acid), are used extensively in dentistry for formulating a family of composites known as glass polyalkenoates, or more commonly glass ionomers (GIs).[1-4] Conventional GIs, specifically, formulations that harden only by an acid–base reaction, consist of an aqueous solution of the aforesaid-type polymers combined with a solid glass, such as a calcium fluoroaluminosilicate (CaFAlSi) powder. To improve the toughness and expand GI uses, the properties of GIs have been improved by development of visible light–curable (VLC) formulations, where the formulations set by both an acid–base reaction and formation of covalent bonding. A brief discussion of the history and current state of art relating to organic polyacid–inorganic composites used in dental restoratives is provided in this chapter.[4] In addition, our past efforts to toughen both the conventional as well as the VLC GI matrix, via the use of N-vinylpyrrolidone (NVP) modification of acrylic acid/itaconic acid copolymers, is presented and discussed. Expansion of the latter type chemistry, to include the design of poly(acrylic acid-*co*-maleic acid-*co*-NVP) materials for formulating conventional GIs, with improved properties, is now described and discussed in this chapter.

II. GLASS IONOMER TECHNOLOGY

Glass ionomer restoratives, which are now used extensively by dentists, were first introduced to the dental materials market about 1971.[5-7] Conventional GIs, namely, formulations, which harden by only an acid–base reaction, consist of an aqueous solution of acrylic acid–based copolymers combined with an acid decomposable glass powder, such as a calcium fluoroaluminosilicate (CaFAlSi).[1-4] Various versions or modification of this type of inorganic glass have been studied;[1,2,8-12] only a few are currently being used. Some of the modifications have included addition of some high-molecular-weight polyacid as well as photoinitiators for VLC-type formulations. The solid/liquid ratios used in the various formulations are critical in determining the handling and final properties observed in the set restorative. Combination of the acidic polymer solution with the glass powder brings about acid hydrolysis of the glass, liberating Al^{3+}, Ca^{2+}, and F^- ions (Fig. 1). In the setting or hardening of the formulations, the calcium and aluminum cations form salt bridges with the ionized carboxylic acid groups pendant on the polymer backbone, as illustrated in Figure 2. The latter brings about hardening of the matrix via ionic or salt-bridge-type crosslinking.

Low levels of d,l-tartaric acid is added to the GI formulations[1,2] to assist in liberating and or facilitating diffusion of the Al^{3+} cations, which are desirable for displacing some of the more readily formed Ca^{2+} salt-bridge cations. The many attractive properties of GIs include such things as anticaries capability, due to fluoride ion liberation, good biocompatibility, low cytotoxicity, excellent adhesion to

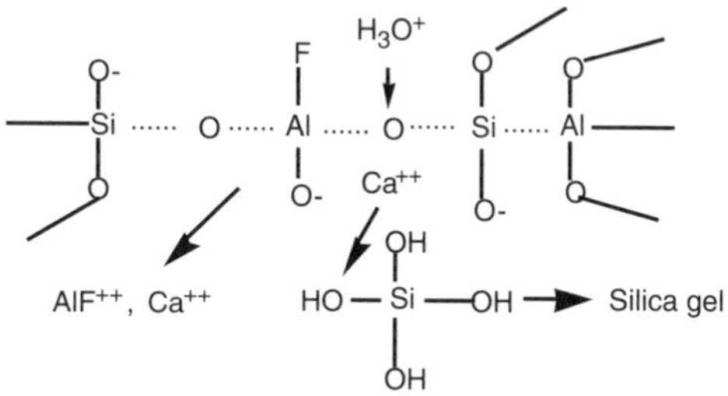

Figure 1 Acid attack and hydrolysis of CaFAlSi glass powder.

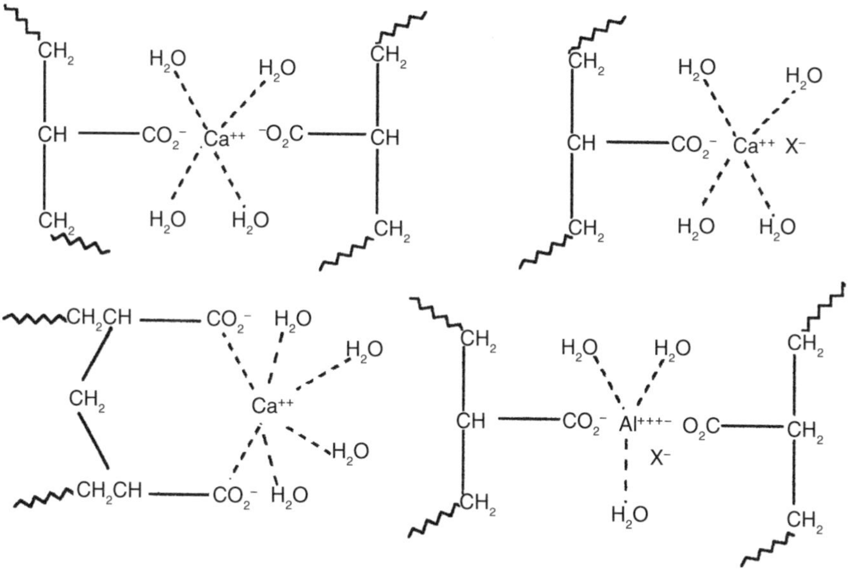

Figure 2 Some possible salt bridges formed in hardened GIs.

tooth surface, and a low coefficient of thermal expansion. Other important advantages include the following: (1) GI restoratives are hard after setting, (2) the hardening process exhibits no exothermic reaction on setting, (3) the compositions show very little to no shrinkage during setting, and (4) the hardened materials have high dimensional stability. Another important consideration in using GIs resides in the fact that conventional type GIs have no free monomer(s) in the matrix, which could leach out into the oral cavity.

One major disadvantage of the GI-type cement is associated with early moisture sensitivity, due to their relatively slow maturation process. Thus, dentists must take great care, immediately after the application, to protect the cement from a wet environment. Other disadvantages include the cement remaining vulnerable to desiccation, showing low resistance to wear and pressure, and having less tensile and flexural strength than do commonly used amalgams and polymeric composites. They also exhibit poor tolerance to acids, which promotes hydrolysis.[1–4] Further, conventional GI formulations suffer from having a modulus of elasticity too low; that is, they exhibit brittle failure, which makes them unsuitable for use as restorations in high-stress areas. To improve the toughness and expand GI uses, the properties of GIs have been modified by development of resin modified glass ionomers or visible light–curable (VLC) formulations, with the effort starting about 1980.[13–21] In the latter case, the acidic copolymers are functionalized with both free-radical polymerizable moieties as well as carboxylic acid entities. VLC GI formulations may also be modified with water-soluble monomers, such as 2-hydroxyethyl methacrylate (HEMA), to improve working requirements for the formulations and assist in gaining improved properties.[20] In the VLC formulations both ionic bonding and covalent bonding provide crosslinking in the hardened matrix, bringing about a significantly improved GI, as pictured or visualized in Figure 3.

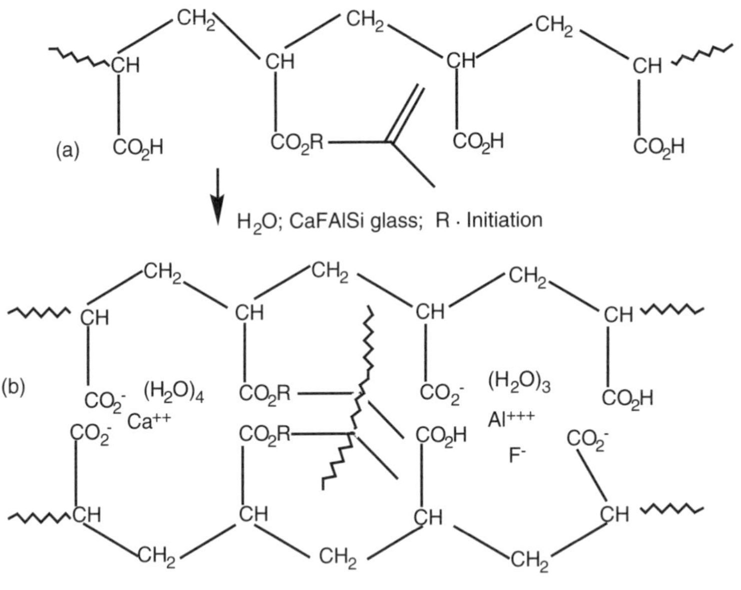

Figure 3 Crosslinking of unsaturated GI (a) via both salt bridge and covalent bonding (b).

As noted earlier, copolymers currently used in conventional GIs are water-soluble polyelectrolytes having carboxylic acid groups directly or very closely attached to the copolymer backbone. The two primary examples are poly(acrylic acid-*co*-itaconic acid) (Fig. 4) or poly(acrylic acid-*co*-maleic acid) (Fig. 5); both are currently used in commercial formulations.

In the process of hardening, due to formation of the ionic crosslinks, a limited number of the carboxylic acid groups are converted to carboxylate-based salt groups. When a certain number of carboxylic acid groups are ionized, the negative charge on the polymer chain increases to such a degree that the remaining carboxylic acid

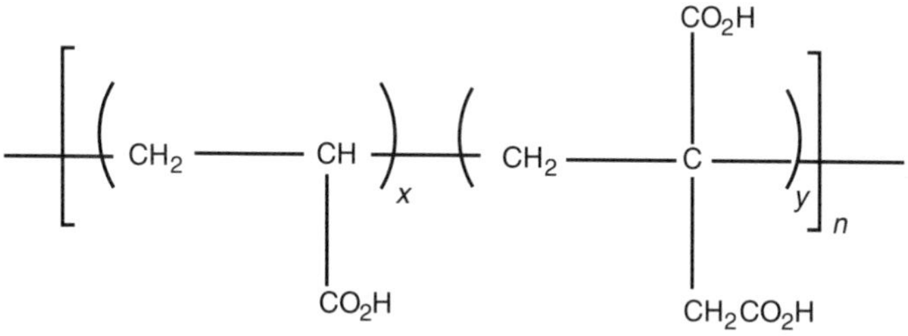

$$\left[\left(CH_2 \underline{\quad\quad} CH\right)_x \left(CH_2 \underline{\quad\quad} \underset{|}{C}\right)_y\right]_n$$

Figure 4 Poly(acrylic acid-*co*-itaconic acid) used in GIs.

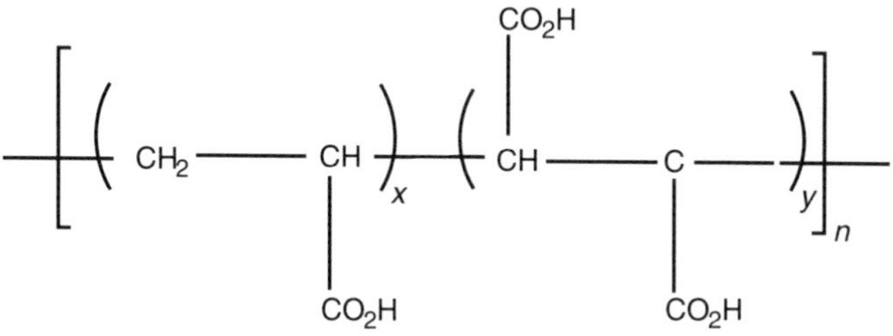

$$\left[\left(CH_2 \underline{\quad\quad} CH\right)_x \left(CH \underline{\quad\quad} C\right)_y\right]_n$$

Figure 5 Structure of poly(acrylic acid-*co*-maleic acid) used in GIs.

groups will not ionize. Hence, they cannot react with the basic glass to form salt bridges.[1] In addition, when the density of crosslinks increases to a certain level, diffusion of Al^{3+} metal ions decrease. As mentioned earlier, tartaric acid is used at low levels in both conventional and VLC GIs, to facilitate Al^{3+} diffusion in the matrix and assist in forming the more desired Al^{3+} based ionic crosslink or salt bridge. Even though tartaric acid is used, very few Al^{3+} tricarboxylate salt bridges (Fig. 6) are actually formed,[1] due to steric effects, making it very difficult for three different polyacid chains to be tied together by an intermolecular salt bridge or crosslink. In other words, steric hindrance, because the carboxylic acid groups are closely attached or directly attached to the copolymer backbone, along with lack of polymer mobility after hardening, play major roles in preventing Al^{3+} di- and tricarboxylate salt-bridge formation.

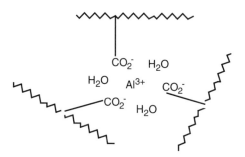

Figure 6 Intermolecular Al tricarboxylate salt bridge.

For some time our group has been exploring ways to improve both conventional and VLC type GIs.[4,22–24] In essence, we believe that the desired improvement of strength and wear resistance of both conventional and VLC-type GIs cannot be achieved with the current copolymers presently being used, due to inadequate Al^{3+} salt bridge formation in the set cements and the lack of copolymer toughness. As a result, our research has focused in the past on making the carboxylic acid groups more available for Al^{3+} salt-bridge formation and finding ways to promote copolymer structural disorder. Presumably, introduction of more structural disorder would help to make the acid groups more available for salt bridge formation, as well as reduce the viscosity of the aqueous copolymer solutions. The lower viscosity would allow for the use of higher-molecular-weight copolymers in the formulations, which in turn would improve the GI toughness and wear resistance.

Attention is now focused on more recent glass ionomer systems or formulations we have studied. In particular, we briefly discuss studies and findings centered on preparing and evaluating amino acid and *N*-vinylpyrrolidone (NVP) modified polyelectrolytes and their evaluation in glass ionomers.

A. Amino Acid–Modified Glass Ionomers

Early efforts to explore our concepts centered on preparing and evaluating a family of copolymers based on acrylic acid–, itaconic acid–, and amino acid–functionalized monomers having acrylamide or methacrylamide type double bonds.[22–48] The amino acid–based monomer used to modify the polyelectrolytes, which was most explored, was based on the glutamic acid derivative shown in Figure 7.

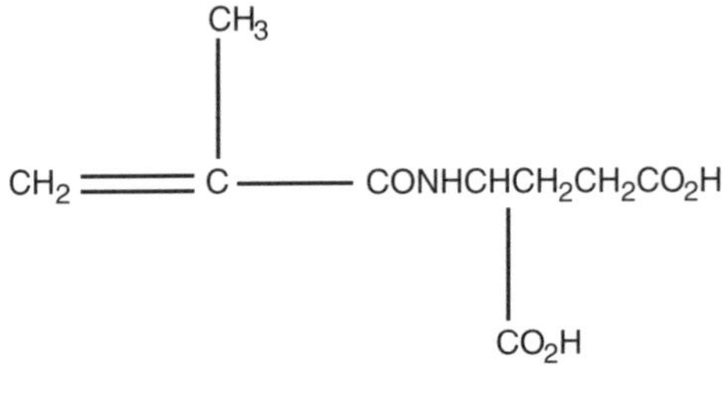

$$CH_2 = \overset{\overset{\displaystyle CH_3}{|}}{\underset{\underset{\displaystyle CO_2H}{|}}{C}} - CONHCHCH_2CH_2CO_2H$$

Figure 7 *N*-Methacryloylglutamic acid (MGA) monomer studied in glass ionomers.

In the amino acid–functionalized polyelectrolytes study, it was reasoned that MGA type monomers would significantly enhance the availability of carboxylate groups for salt bridge formation, as well as promote a certain level of disorder in the copolymer backbone. The latter assumption follows documentation[49–52] that reactive functionalities removed from direct attachment to a polymer backbone, via the use of spacer groups, allows greater access for reactions, especially on a rigid polymer backbone. The spacer moiety or tethering unit serves to dissociate steric hindrance of the polymer mainchain from the assembly, while not affecting the reactivity of the spacer linked functionality. The functionality, in this case a $-CO_2H$ group, would have more freedom to form a salt bridge. The tethered or spacer linked functionality has much greater degrees of freedom from constraint by the polymer backbone since the rigid polymer chain cannot bend or flex sufficiently to make more of the $-CO_2H$ groups available for reaction. Further, by use of the tethered amino acids some of the $-CO_2H$ groups are prevented from being on adjacent carbons, as in poly(AA-*co*-MA). At the same time, however, the length and type of tethering unit can be utilized to influence changes in hydrophilicity, conformation in solution, viscosity, solubility of the polymers, etc. Thus, it was reasoned that the $-CO_2H$ groups tethered to the polyelectrolyte backbone, with various lengths of spacer moieties, would have much greater degrees of freedom to react with Ca^{2+} and Al^{3+} ions liberated from the reactive glass, resulting in enhanced basic glass–acidic polymer interaction, more complete salt-bridge formation between the $-CO_2^-$ anions and the Al^{3+} cations, and enhanced homogeneity in the matrix. The latter concepts were part of the thoughts that went into the accomplished study. Subsequent research studies proved this to be an accurate assumption since incorporation of the amino acid groups did bring about greater salt-bridge formation as shown by Raman spectroscopy.[25,26] It was also postulated that the approach would help to provide a more uniform release of fluoride ions. Thus, tethering amino acids to the polymer backbone provides more freedom for the acid groups and less steric hindrance, availability of different types of carboxylic acid residues, and different pH values, promoting more salt-bridge formation. This new approach was found to improve GIs. As an example, it was definitely shown that copolymers having $x/y/z$ ($AA_xIA_yMGA_z$) ratios of 7–3–3 and 8–1–1 monomers in the backbone (Fig. 8) functioned in excellent fashion for formulating a conventional GI with improved mechanical properties. For example, the 8–1–1 copolymer based formulation was found optimum for producing a conventional GI having a compressive strength (CS) of 269.9 (24.8) MPa, compared to the commercial Fuji II control having CS = 224.9 (25.8) MPa. In the study, all test samples, including the control, were prepared and conditioned in the same fashion. Also, the experimental formulation used the Fuji II glass at the same P/L ratio recommended for Fuji II. Both systems had comparable hardness, or KHN values, with the control and experimental being, respectively, 36.7 and 36.5. The experimental 7–3–3 copolymer formulation compared to the Fuji II control, using the same Fuji II glass powder

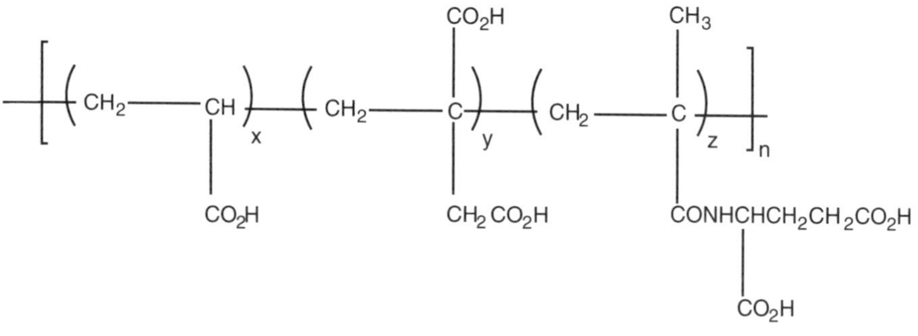

Figure 8 Poly(AA-*co*-IA-*co*-MGA) studied in glass ionomers.

at the same P/L ratio, had a flexural strength (FS) of 34.6 (7.0) MPa and fracture toughness (FT) of 0.64 (0.13) Mn/m $^{1.5}$, compared to Fuji II, having FS = 14.7 (0.7) MPa and FT = 0.48 (0.08) Mn/m $^{1.5}$.

We also focused on producing and evaluating resin modified or hybrid, light-cured (VLC) GIs, based on the same polyelectrolytes having tethered amino acid residues.[27] As an example, the 8–1–1 poly(AA-co-IA-co-MGA) was treated with 2-isocyanatoethyl methacrylate (IEM), using a 3M-patented[20] procedure, to convert approximately 15% of the carboxylic acid residues into grafted IEM segments. Figure 9 provides a picture of one possible structure for the IEM treated poly(AA-co-IA-co-MGA). The purified IEM grafted copolymer (45 parts) was blended with 21 parts of HEMA and 34 parts of water, along with a suitable photoinitiator system. After mixing the copolymer solution with the Fuji II LC glass powder, at the same P/L ratio recommended for the control (Fuji II LC), the experimental VLC formulation was cured with visible light, conditioned and evaluated alongside the Fuji II LC control samples, which were prepared and conditioned in the same way as the experimental samples. It was found that the experimental VLC GI exhibited CS = 318.7 (5.5) MPa, versus Fuji II LC having CS = 283.0 (41.7) MPa. Further, using design of experiments (DOE) or statistical methods (SAS),[27] it was found that a formulation having 52.5 parts of IEM grafted copolymer, 18 parts of HEMA, and 29.5 parts of water, formulated with the Fuji II LC glass powder at the same P/L ratio used in Fuji II LC, exhibited CS = 370.3 (8.8) MPa. Here again we point out that Raman spectroscopy showed greater salt-bridge formation occurred in the experimental samples compared to the control.[25,26] Clearly, amino acid modification of polyelectrolytes having carboxylic acid groups tethered various distances of the copolymer backbone as illustrated by using MGA (Figs. 7 and 8), can be used to prepare improved conventional as well as VLC GIs.

In another study using MGA, various commercial and experimental VLC GI formulations were modified by addition of small amounts of MGA to the polymer solutions used to prepare the VLC GI test samples. In this approach, the MGA underwent free-radical copolymerization within the organic matrix. In the effort, it was shown that MGA added to the formulations significantly enhanced fluoride ion released, both in quantity as well as over time, when compared to test samples made without MGA.[40] Further MGA addition improved the hardness of the materials as well as adhesion to tooth structure.[37,38,44,45]

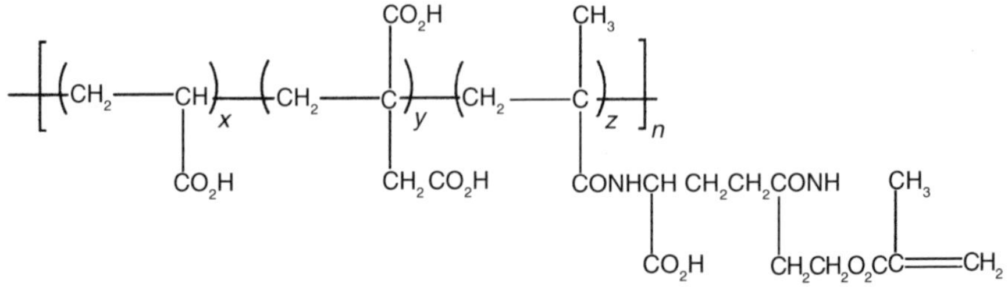

Figure 9 Possible structure of IEM-treated poly(AA-co-IA-co-MGA).

B. *N*-Vinylpyrrolidone (NVP)-Modified Glass Ionomers

In more recent studies, we started to look at ways to incorporate flexible spacers within the backbone of copolymers to be used for formulating GIs, while maintaining water solubility for the copolymers as well as allowing for a suitable number of carboxylic acid groups to be available for salt-bridge formation. To do this, we started looking at NVP modified GIs,[53–63] seeking to enhance salt-bridge formation by increasing the flexibility and disorder in the backbone, as well as foster greater ionization of the acid groups. We hypothesized that NVP modification of the acrylic acid/itaconic acid copolymers, as shown in Figure 10, could improve both the water retention characteristics of the GI and the working characteristics for the formulation, as well as possibly enhance GI mechanical properties.

The nonionogenic synthetic NVP polymer is rather hygroscopic and readily soluble in water. Because of this strong interaction with water or hydrophilicity, as well as nontoxicity and high proclivity for complexation, NVP containing polymers have been extensively explored and used in the medical, food, textile, pharmaceutical, and biomaterial industries. Thus it was felt that there would be no adverse thoughts about having NVP in a copolymer backbone used in the oral cavity. As mentioned, NVP was chosen because it can cause further disorder or regularity of the copolymer chain, possibly providing more opportunity for conformational transition and rotation of polymer molecules, as suggested from studies made by others on copolymers containing maleic acid and NVP.[64] NVP absorbs water because of strong hydrogen bonding, which could possibly lead to a longer working time for GI formulations. NVP should also bring about some interruption of the strong hydrogen bonds between various carboxylic acid groups. Thus, incorporation of a low level of NVP in the copolymer backbone could possibly facilitate lowering the viscosity of the copolymer solutions. It is apparent that elimination of some AA and IA or MA monomers in the copolymer backbones is possible, since few of the acid groups on any of these copolymers are actually part of salt bridges, especially Al^{3+}-based salt bridges.[1,25,26]

Early work showed that poly(AA-*co*-IA-*co*-NVP) materials (Fig. 10), having low levels of NVP in the copolymer backbone, could be utilized to prepared GIs having improved microhardness, flexural strength, fracture toughness, and improved adhesion to tooth structure.[53–55,57,58] Design of experiments (DOE) and SAS statistical

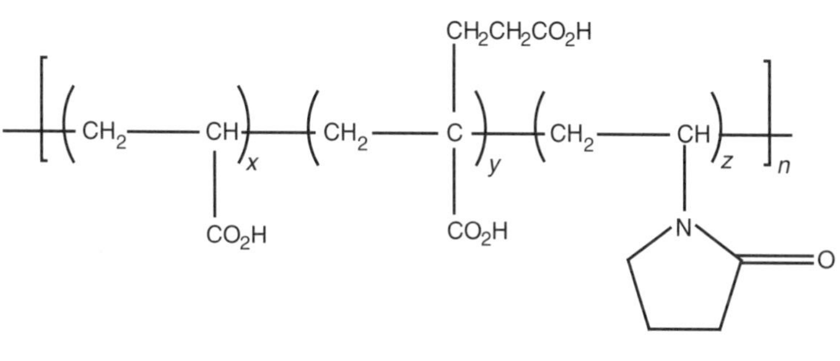

Figure 10 Poly(AA-*co*-IA-*co*-NVP) studied in glass ionomers.

methods[53,58] were used to optimize molecular weights (MW) of the NVP copolymers as well as the level of NVP present in the polyelectrolyte backbone. The optimal molar ratio of the copolymers composed of AA, IA, and NVP was 7–1–3, respectively, as shown by SAS statistical efforts, with the focus on flexural strength (FS) measurements. As an example, the 7–1–3 copolymer, having a number-average molecular weight of 10,800, as measured by GPC, exhibited an 85% increase of FS compared to a typical and currently used conventional GI. The work clearly showed that FS increased with copolymer molecular weight. However, in a two-part formulation, there is a limit on how high the copolymer MWs may be, since the dental technician must be able to adequately mix, control working and setting time, or blend the aqueous solution of the copolymer with the glass powder.

Poly(AA-*co*-IA-*co*-NVP) materials (Fig. 10), having different MWs and NVP in the copolymer backbone, were treated with IEM (Fig. 11) to prepare VLC GIs.[54,58] Again, using DOE and SAS methods,[58] it was determined that an 8–2–1 (AA–IA–NVP) copolymer worked best to prepare light-curable (VLC) glass ionomer cements having higher compressive and flexural strength values than well-known commercial products. Clearly, the aforesaid findings suggest that NVP could have a place in the design of improved GIs. With the latter in mind we describe, in the rest of the text, our very recent efforts to modify poly(acrylic acid-*co*-maleic acid) with NVP, as a route to prepare improved, conventional GIs.

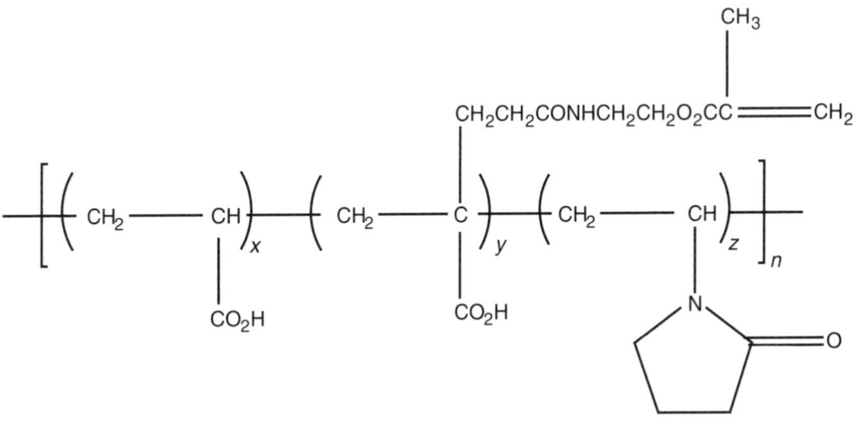

Figure 11 One possible structure for IEM-treated poly(AA-*co*-IA-*co*-NVP) studied in GIs.

III. NEW NVP-MODIFIED GLASS IONOMERS: EXPERIMENTAL WORK

A. Materials

Acrylic acid (AA), maleic acid (MA), potassium persulfate, and solvents were used as received from Aldrich Chemical Co. The *N*-vinylpyrrolidone (NVP) was distilled prior to use. The CaFAlSi glass powders, used in preparing the experimental formulations, were part of commercial glass ionomer kits: Fuji II and Fuji IX, supplied by GC America Corp.

B. Polymer Synthesis

A mixture of potassium persulfate (0.94 g) in 260 mL of distilled water was stirred under a nitrogen atmosphere for 30 min to remove dissolved oxygen. A mixture of AA (65.58 g, 0.91 mol), MA (45.27 g, 0.39 mol), NVP (14.44 g, 0.13 mol), and 130 mL of distilled water was prepared. An initiator solution consisting of $K_2S_2O_8$ (0.94 g) in 80 mL of distilled water was also prepared. Both mixtures were subdivided into 20 equal portions and subsequently added every 5 min to a reactor, where the temperature was regulated at 95°C. The polymerization reaction was carried out for 24 h, followed by isolation of the terpolymer by a standard freeze-drying technique, using Edwards High Vacuum International (Sussex, UK) equipment. The terpolymer was dissolved in methanol and combined slowly with a large volume of ethyl acetate to precipitate the purified material. After collection, the white copolymer powder was obtained in a 95 g (76%) yield. The structure of the copolymer (Fig. 12) was confirmed by FTIR and NMR (^{1}H and ^{13}C) spectroscopy.[54,58]

C. Characterization

FTIR spectra were obtained on a MIDAC spectrophotometer. The ^{1}H NMR and ^{13}C NMR spectra were collected on a 300-MHz Bruker AM spectrometer, using deuterated water as a solvent and with trimethylsilane (TMS) as internal standard. Viscosities of the copolymer solutions (50–50, wt–wt) were measure by a cone and plate viscometer, using a Cap 2000 Viscometer (Brookfield Engineering Laboratories, Inc., Stoughton, MA), at 25°C. An Instron universal testing machine (Instron Model 4204, Instron Corp., Canton, MA) was used to determine mechanical properties.

D. Physical Properties

All copolymers were dissolved in distilled water (50–50, wt–wt), for forming the viscous solutions used in the formulation of experimental GIs. The Fuji II and Fuji IX glass powders were selected for blending with the copolymer solutions, with a powder/liquid (P/L) of 2.7/1 and 3.2/1, respectively. The Fuji II and Fuji IX systems were also blended and allowed to harden, using the manufacture recommended P/L ratio. This provided both a control and experimental GI system, prepared under the same conditions, for determination of the mechanical properties relative to each other. Six specimens for both the experimental and Fuji II and Fuji IX were prepared. Properties tested were compressive strength (CS), diametral tensile strength (DTS), and flexural strength (FS), using test specimen sizes, respectively, of 4×8 mm, 4×2 mm and $2\times2\times25$ mm. All samples were conditioned in distilled water at 37°C for 7 days before being tested.

IV RESULTS AND DISCUSSION

As described above, NVP was copolymerized with acrylic acid (AA) and maleic acid (MA) in various molar ratios, producing poly(AA-*co*-MA-*co*-NVP) (Fig. 12), for examination as a polyelectrolyte for formulation of improved, conventional GIs (Table 1). Any improvement of conventional GIs could be very important, since this type of GI would be utilized to a much greater degree in population areas where visible light curing is not available. Incorporation of an optimal molar ratio of NVP in the polyelectrolyte, where NVP provides a strong water sorption center and bulky structure, may prolong the acid–base reaction, break up the rigidity of the polymer chain, and increase accessibility of carboxyl groups to form complexes with metal cations. providing an improved setting reaction and physical properties.[62] As shown in Table 1, copolymers having acrylic acid, maleic acid, and NVP residues were prepared in excellent yields, with the copolymers having several AA–MA–NVP molar ratios. The molar ratios, viscosities (Vis), yield stresses (YS), and confidences to fit (CoF) are also outlined in Table 1. The molecular weights of the copolymers were intentionally varied, as shown by the viscosity measurements in Table 1. Further, the molecular weights were kept low, allowing direct comparison of the NVP containing polyelectrolytes with the copolymers used in Fuji II and Fuji IX conventional GIs.

To prepare samples for mechanical properties testing, the aqueous copolymer solutions (50/50 wt/wt) in Table 1 were formulated with the CaFAlSi glass powders used in the two commercial GIs: Fuji II and Fuji IX. In the blending of the solid glass powder with the aqueous solution of the copolymers, at the same P/L ratios recommended for Fuji II and Fuji IX, it was observed that the presence of the NVP allowed generation of a more smooth blend compared to the two commercial materials. Mechanical properties of the experimental materials were compared to the Fuji II and Fuji IX controls, where all test samples ($n = 6$) were prepared and conditioned under the same procedure. The flexural strengths (FS), flexural moduli (FM), compressive strengths (CS), and compressive moduli (CM) of the experimental copolymers were compared to the same Fuji II and Fuji IX mechanical properties, as shown

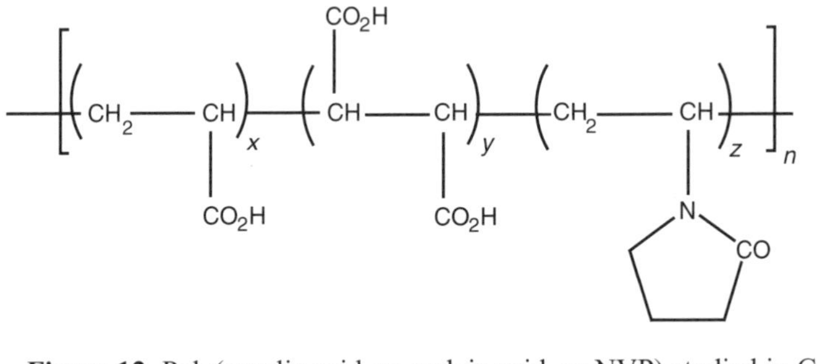

Figure 12 Poly(acrylic acid-*co*-maleic acid-*co*-NVP) studied in GIs.

in Tables 2 and 3. The results show that there are several molar ratios and molecular weight options to focus on for optimizing and obtaining the desired improvements in conventional GIs.

V. CONCLUSIONS

It was demonstrated that the NVP monomer is a viable candidate to study for improvement of conventional glass ionomers, especially when used at low levels in the poly(AA-*co*-IA-*co*-NVP) or poly(AA-*co*-MA-*co*-NVP)-type polyelectrolytes. For example, if an improvement of the flexural strength or compressive strength is desired, copolymers A and F in Tables 1 and 2 could possibly be useful. Further, the results in Tables 2 and 3 clearly show that all NVP containing polyelectrolytes improve the flexural strength of GIs compared to the best commercial GIs. In any case, statistical design of experiments would need to be used to find the most desirable molar ratios of monomers to use in the poly(AA-*co*-MA-*co*-NVP), as well as the highest molecular weight of copolymer that could be used in blending the glass powder with the aqueous polymer solution. It would also be useful to look at variations in the composition of the glass powder used in the formulations, as well as obtain the optimal powder/liquid ratio to use in any particular formulation. Another area needing study is concerned with knowing and varying the microstructure of the copolymers, along with evaluating how the latter structural differences influence GI mechanical properties.

Table 1 Poly(AA-*co*-MA-*co*-NVP) Compositions and Viscosities

System	AA–MA–NVP Molar Ratios	Viscosity (SD)[a] (cP)	YS (SD) N/M^2
Fuji II	——	562 (21)	44 (17)
Fuji IX	——	565 (7)	43 (11)
A	5–5–1	730 (51)	35 (22)
B	5–5–1	1820 (5)	63 (11)
C	7–1–3	526 (52)	12 (10)
D	7–3–1	1109 (56)	34 (21)
E	7–3–1	1732 (36)	103 (25)
F	7–3–1	1946 (42)	51 (39)
G	8–1–1	894 (44)	317 (40)

[a]Viscosities determined on a Cap 2000 Viscometer (Brookfield Engineering), at 25°C.

Table 2 Mechanical Properties of NVP Copolymers Formulated with Fuji II Glass, Compared to Fuji II

System[a]	FS (SD) (MPa)	CS (SD) (MPa)
Fuji II	11.9 (1.3)	249 (6)
A	35.5 (7.6)	256 (16)
B	44.0 (2.1)	205 (19)
C	37.0 (4.9)	188 (13)
D	39.0 (4.7)	220 (20)
E	41.5 (4.1)	217 (18)
F	39.8 (4.8)	242 (10)
G	38.7 (3.8)	239 (4)

[a]The powder/liquid (P/L) ratios used in all formulations were the same as for Fuji II, namley, 2.7/1. All samples were conditioned in distilled water at 37°C for 7 days.

Table 3 Mechanical Properties of NVP Copolymers Formulated with Fuji IX Glass, Compared to Fuji IX

System[a]	FS (SD) (MPa)	CS (SD) (MPa)
Fuji IX	16.7 (1.9)[b]	281 (19)[c]
A	40.5 (4.6)	290 (21)
B	48.6 (3.5)[b]	267 (13)[c]
C	36.0 (6.0)	186 (23)
D	32.1 (5.4)	231 (16)
E	39.4 (3.4)	265 (22)
F	30.1 (3.4)	307 (28)
G	38.7 (3.6)	242 (15)

[a]The powder/liquid (P/L) ratio used was the same as for Fuji IX, namely, 3.2/1. After preparation and before testing, all samples were conditioned in distilled water at 37°C for 7 days.
[b]Flex moduli (FMs) for Fuji IX and B were, respectively, 16.5(0.9) and 20.2(0.4) GPa.
[c]Compressive moduli (CMs) for Fuji IX and B were, respectively, 6.90(0.23) and 6.64(0.12) GPa.

VI. REFERENCES

1. A. D. Wilson and J. W. McLean, *Glass-Ionomer Cement*, Quintessence Publishers, Chicago, 1988.
2. A. D. Wilson and J. W. Nicholson, *Acid Base Cements; Their Biomedical and Industrial Applications*, Cambridge Univ. Press, Cambridge, UK, 1993.

3. J. W. McLean, J. W. Nicholson, A. D. Wilson, *Quint. Int.* **25**(9), 587 (1994).

4. B. M. Culbertson, "Glass-Ionomer Dental Restoratives," *Progr. Polym. Sci.* **26**, 577–604 (2001).

5. C. D. Smith, *J. Can. Dent. Assoc.* **37**, 22 (1971).

6   A. D. Wilson, B. E. Kent, *J. Appl. Chem. Biotechnol.* **21**, 313 (1971).

7. A. D. Wilson, B. E. Kent, *Br. Dent. J.* **1321**, 133 (1972).

8. S. Crisp, A. J. Ferner, B. G. Lewis, A. D. Wilson, *J. Dent.* **3**, 125 (1975).

9. J. W. McLean, A. D. Wilson, *Austral. Dent. J.* **22**, 31 (1977).

10. B. E. Kent, B. G. Lewis, A. D. Wilson, *J. Dent. Res.* **58**, 1607 (1980).

11. S. Crisp, B. E. Kent, B. G. Lewis, A. J. Ferner, A. D. Wilson, *J. Dent. Res.* **59**, 1055 (1980).

12. A. D. Wilson, H. J. Prosser, *Br. Dent. J.* **157**, 449 (1984).

13. H. Norbo, *Glass-Ionomers Cementes*, 3M Dental Bulletin 1, 1990.

14. T. P. Croll, C. M. Killian, *Quint. Int.* **23**, 679 (1991).

15. T. P. Croll, *Quint. Int.* **22**, 137 (1991).

16. A. W. Bourke, A. W. Walls, J. F. McCabe, *J. Dent.* **20**, 115 (1992).

17. T. P. Croll, *Quint. Int.* **24**, 109 (1993).

18. G. M. Knight, *Quint. Int.* **25**, 97 (1994).

19. J. M. Antonucci. J. M. McKinney, J. W. Stansbury, U.S. Patent 7,160,865 (1988).

20. S. B. Mitra, U.S. Patent 5,130,347 (July 14, 1992); S. B. Mitra, S. Mitra, J. M. Gartner, B. L. Kedrowski, *ACS Polym. Preprints* **35**(2), 77 (1994).

21. D. C. Smith, *J. Am. Dent. Assoc.* **120**, 20 (1990).

22. B. M. Culbertson, E. C. Kao, U.S. Patent 5,369,142 (1994).

23. E. C. Kao, B. M. Culbertson, D. Xie, *Dent. Mater.* **12**, 44–51 (1996).

24. B. M. Culbertson, E. C. Kao, D. Xie, *ACS Polym. Mater. Sci. Eng.* **71**, 520–523 (1994).

25. Z. Ouyang, S. K. Sneckenberger, E. C. Kao, B. M. Culbertson, P. W. Jagodzinski, *Appl. Spectrosc.* **53**(3), 297–301 (1999).

26. E. C. Kao, Z. Ouyang, S. Snecenberger, D. Xie, P. W. Jagodzinsi, B. M. Culbertson, *J. Dent. Res.* **77A** 169 (1998) (Abst. 512).

27. M. H. Dotrong, W. M. Johnston, B. M. Culbertson, *J. Macromol. Sci. Part A, Pure & Appl. Chem.* **A37**(8), 911–926 (2000).

28. B. M. Culbertson, M. H. Dotrong, *J. Macromol. Sci. Part A, Pure & Appl. Chem.* **A37**(5), 419–431 (2000).

29. D. Xie, W. A. Brantley, B. M. Culbertson, G. Wang, *J. Dent. Mater.* **16**(2), 129–138 (2000).

30. B. M. Culbertson, D. Xie, A. Thakur, *J. Macromol. Sci. Part A, Pure & Appl. Chem.* **A36**(5/6), 681–696 (1999).

31. B. M. Culbertson, D. Xie, A. Thakur, *Macromol. Symp.* **131**, 11–18 (1998).

32. B. M. Culbertson, D. Xie, *ACS Polym. Preprints* **39**(2), 765–766 (1998).

33. B. M. Culbertson, A. Thakur, D. Xie, E. C. Kao, *ACS Polym. Preprints* **38**(2): 127–128 (1997).

34. B. M. Culbertson, *ACS Polym. Mater. Sci. Eng. Preprints* **71**, 520–522 (1994).

35. B. M. Culbertson, E. C. Kao, D. Xie, A. Thakur, *ACS Polym. Symp.*, *MakroAkron* **94**, 512 (1994).

36. M. H. Dotrong, B. M. Culbertson, *J. Dent. Res.* **80**, 252 (2001) (Abst. 1730).

37. B. M. Culbertson, M. H. Dotrong, *J. Dent. Res.* **79** (2000) (Abst. 372).

38. E. C. Kao, B. M. Culbertson, A. Thakur, W. M. Johnston, *J. Dent. Res.* **76**, 75 (1997) (Abst. 492).

39. L. D. Knebel, B. M. Culbertson, A. Thakur, *J. Dent. Res.* **76**, 75 (1997) (Abst. 496).

40. B. M. Culbertson, A. Thakur, *J. Dent. Res.* **76**, 76 (1997) (Abst. 497).

41. E. C. Kao, S. Sharma, K. Hosteler, B. M. Culbertson, G. Hobbs, *J. Dent. Res.* **76**, 317 (1997) (Abst. 1208).

42. B. M. Culbertson, D. Xie, J. Xu, *J. Dent. Res.* **75**, 317 (1996) (Abst. 2429).

43. B. M. Culbertson, A. Thakur , D. Xie, E. C. Kao, *J. Dent. Res.* **75**, 292 (1996) (Abst. 2200).

44. L. D. Knebel, B. M. Culbertson, A. Thakur, W. M. Johnston, *J. Dent. Res.* **75**, 293 (1996) (Abst. 2202).

45. L. D. Knebel, B. M. Culbertson, A. Thakur, R. E. Kerby, *J. Dent. Res.* **74**, 38 (1995) (Abst. 211).

46. E. C. Kao, D. Xie, B. M. Culbertson, W. M. Johnston, *J. Dent. Res.* **74**, 106 (1995) (Abst. 755).

47. A. Thakur, B. M. Culbertson, *J. Dent. Res.* **74**, 106 (1995) (Abst. 756).

48. B. M. Culbertson, E. C. Kao, D. Xie, *J. Dent. Res.* **72**, 385 (1993) (Abst. 2256).

49. B. M. Culbertson, E. C. Kao, D. Xie, *J. Dent. Res.* **72**, 384 (1993) (Abst. 2250).

50. H. Finklemen, H. Ringsdorf, J. H. Wendorff, *Makromol. Chem.* **179**, 273 (1978).

51. R. Barbucci, M. Casolaro, A. Magnai, *Makromol. Chem.* **190**, 2627–2638 (1989).

52. R. Barbucci, M. Casolaro, A. Magnai, C. Roncalini, *Macromolecules* **24**, 1249–1252 (1991).

53. T. Hayahawa, K. Horie, *J. Dent. Mater.* **10**(2), 165–171 (1991).

54. D. Xie, B. M. Culbertson, W. M. Johnston, *ACS Polym. Preprints* **39**(1), 641–642 (1998).

55. D. Xie, B. M. Culbertson, W. M. Johnston, *J. Macromol. Sci. Part A, Pure & Appl. Chem.* **A35**(10), 1631–1650 (1998).

56. D. Xie, B. M. Culbertson, W. M. Johnston, *J. Macromol. Sci. Part A, Pure & Appl. Chem.* **A35**(10), 1615–1629 (1998).

57. B. M. Culbertson, D. Xie, *ACS Polym. Preprints* **39**(2), 765–766 (1998).

58. D. Xie, B. M. Culbertson, G. Wang, *J. Macromol. Sci. Part A, Pure & Appl. Chem.* **A35**(4), 547–561 (1998).

59. B. M. Culbertson, D. Xie, W. M. Johnston, *ACS Symp. Series Publications 755*, K. O. Havelka, ed., March 2000.

60. B. M. Culbertson, D. Xie, *ACS Div. Polym. Chem., Polym. Preprints* **39**(2), 765 (1998).

61. M. H. Dotrong, B. M. Culbertson, *J. Dent. Res.* **80**, 251 (2001) (Abst. 1727).

62. E. C. Kao, Y. Jin, P. W. Jagodzinski, B. M. Culbertson, *Br. Med. J. Dent. Res.* **80**, 251 (2001) (Abst. 1728).

63. D. Xie, B. M. Culbertson, G. Wang, *J. Dent. Res.* **77A**, 170 (1998) (Abst. 513).

64. D. Xie, B. M. Culbertson, *J. Dent. Res.* **76**, 74 (1997) (Abst. 487).

65. H. Rios, L. Gargallo, D. Radic, *J. Polym. Sci., Part B, Polym. Phys.* **24**, 2421–2431 (1986).

66. S. B. Mitra, B. L. Kedrowski, *ACS Polym. Preprints* **38**(2), 129 (1997).

Index

A

Absorption spectra, ferrocene-DNA conjugates, 30–34

Acid attack and hydrolysis, glass ionomer technology, 194–198

AC impedance spectroscopy (ACIS), metallated DNA conjugates, 40–41

Acrylic acid (AA), *N*-vinylpyrrolidone (NVP)-modified glass ionomers, 202–203

Adenine, non-hydrogen-bonding basepairs, 48–49

Alkyltin, polymer anticancer activity, 68–70

Aluminum compounds, glass ionomer technology, 197–198

Amide-linked ferrocene, polymer conjugates, 102–109

Amidomalonic acid (ama) group, mixed platinum-oxygen/platinum-nitrogen bound polymers, 180–182

Amine derivatives:
  biofissionable platinum-nitrogen complexes, 149–154
  cisplatin polymeric compounds, mainchain structures, 137–141
  platinum-oxygen-bound polymers, 173–180

Amino acid derivatives:
  cisplatin mainchain polymers, 141–143
  glass ionomer technology, 198–200
  platinum-oxygen-bound polymers, 179–180
  polymeric platinum-containing anticancer compounds, 183

Ampicillin, organotin polymers, anticancer activity, 66–70

Anchoring mechanisms, biofissionable platinum-nitrogen complexes, 154–161

Androgen receptor (AR), organotin monomers, anticancer activity, 63–65

Anhydride groups, platinum-oxygen-bound polymers, 176–180

Antibody research, polymeric platinum-containing anticancer compounds, 183

Anticancer compounds:

ferrocene-ferricenium systems, 95–98

ferrocene polymer-drug conjugation, 98–100

polymeric platinum-containing drugs:
  basic properties, 120–121
  carrier-bound complexes, nitrogen donor ligands, 147–161
    biofissionable complexes, amine anchors, 149–161
    platinum-polyphosphazenes, 147–149
  cisplatin properties, 127–130
  conjugation strategy, 133–137
  formation mechanisms, 121–125
  future applications, 182–184
  mainchain-incorporated *cis*-diamine-coordinated platinum, 137–147
    amine derivatives, 137–141
    amino acid derivatives, 141–143
    antiviral activity, 145–147
    nitrogen-platinum products, 143–144
    solution stability, 144–145
    thermal stability, 145
  mixed platinum-oxygen/platinum-nitrogen-bound polymers, 180–182
  monomeric structures, 125–126
  nomenclature, 125
  platinum-oxygen-bound polymers, 161–180
  structure-activity relationships, 130–133
tin compounds:
  chelating agents, 59–62
  monomeric molecular studies, 62–65
  polymer activity, 65–70
  research background, 58–59

Antifouling marine coatings, tin-containing biocides, 5–6

Antimony, condensation polymers, 10

Antiproliferative agents:
  cisplatin structure-activity relationships, 130–133
  ferrocene conjugates:
    amide-linked ferrocenes, 102–109
    basic properties, 90–91
    bioactivity screening, 110–113

carrier components, 101–102
ester-linked conjugates, 109–110
ferrocene-ferricenium system, 92–98
future research issues, 113–114
pharmaceutical applications, 98–100
synthesis and structure, 100–110
Antiviral activity, cisplatin mainchain structures, 145–147
AP research compounds, mixed platinum-oxygen/platinum-nitrogen bound polymers, 180–182
Aqueous solutions:
  biofissionable platinum-nitrogen complex anchoring, 156–161
  cisplatin properties, 127–130
  platinum-oxygen-bound polymers, 163–180
Arsenic, biomedical applications, 3
Artificial metallo-DNA, future applications, 54
Avidin-biotin coupling, platinum-oxygen-bound polymers, 164–180

B

Backbone structure:
  amide-linked ferrocene-polymer conjugates, 107–109
  biofissionable platinum-nitrogen complexes, 150–154
  glass ionomer technology, 198–200
  N-vinylpyrrolidone (NVP)-modified glass ionomers, 201–202
  platinum-containing polymer-drug conjugation, 134–135
  polymer-ferrocene conjugates, 101–102
Bacteria:
  organotin degradation, 58–59
  polymeric platinum-containing anticancer compounds, 183–184
Balb/3T3 cells:
  cisplatin mainchain structures:
    amine derivatives, 138–141
    amino acid derivatives, 143
  organotin polymer anticancer activity, 66–70
Bioactive agents:
  amide-linked ferrocene-polymer conjugates, 106–109
  ferrocene-ferricenium systems, 92–98
  organometallics, 4–7
    ferrocene, polymeric systems, 6
    osmarins, 7
    tin-containing biocidal polymers, 5–6
  organotin polymers, anticancer activity, 65–70
  platinum-containing polymer-drug conjugation, 135–136
  platinum-oxygen-bound polymers, 163–180

Bioactivity screening, polymeric ferrocene conjugates, 110–113
Bioassays, amide-linked ferrocene-polymer conjugates, 106–109
Biocidal polymers, tin compounds, 5–6
Biofissionable platinum-nitrogen complexes, primary/secondary amine anchors, 149–154
B 16 melanoma cell lines, mixed platinum-oxygen/platinum-nitrogen bound polymers, 181–182
Butler-Volmer method, ferrocene-DNA conjugates, 32–34
Butyltin species:
  polymer anticancer activity, 69–70
  toxicity, 58

C

Cage structures, small-molecule analogs, biomedical applications, 15–16
Calcium:
  metallated DNA, 37–41
  organotin compound toxicity, 63–65
Calcium fluoroaluminosilicate (CaFAlSi):
  glass ionomer technology, 194–198
  N-vinylpyrrolidone (NVP)-modified glass ionomers, 202–203
Carboplatin, currently approved compounds, 126
Carboxymethyldextran, platinum-oxygen-bound polymers, 164–180
Carrier-bound systems:
  biofissionable platinum-nitrogen complexes, 149–154
  platinum complexes:
    biofissionable complexes, amine anchors, 149–161
    nitrogen donor ligands, 147–161
    platinum-polyphosphazenes, 147–149
  platinum-containing polymer-drug conjugation strategy, 134–137
  platinum-oxygen-bound polymers, 161–180
  polymer-ferrocene conjugates, 101–102
Cell lines. *See also* specific cell lines
  biofissionable platinum-nitrogen complexes, 152–154
  cisplatin mainchain structures, amine derivatives, 137–141
  ferrocene-ferricenium systems, biological activity, 96–98
  organotin cytotoxicity, 63–65
  polymeric ferrocene conjugates, 111–113
Cephalexin, organotin polymer anticancer activity, 67–70
CH1 cell line, ferrocene-ferricenium systems, biological activity, 96–98

Chelation:
   biofissionable platinum-nitrogen complexes, 154
   cisplatin structure-activity relationships, 130–133
   mixed platinum-oxygen/platinum-nitrogen bound polymers, 180–182
   organotin compounds, 59–62
Chitosan, biofissionable platinum-nitrogen complexes, 152–154
Chromosome structures, organotin monomers, anticancer activity, 63–65
cis-Aq compound, platinum-oxygen-bound polymers, 179–180
Cisplatin:
   basic properties, 127–130
   currently approved compounds, 125–126
   mainchain structures, 137–147
      amine derivatives, 137–141
      amino acid derivatives, 141–143
      antiviral activity, 145–147
      nitrogen-platinum products, 143–144
      solution stability, 144–145
      thermal stability, 145
   organotin anticancer activity, 60–62
   platinum-oxygen-bound polymers, 163–180
   polymer-drug conjugation strategy, 133–137
   polymeric ferrocene conjugates *vs.*, 111–113
   research background, 120–121
Cobalt:
   metallated DNA, 37–41
   small-molecule analogs, biomedical applications, 12–16
Colo 320 DM, polymeric ferrocene conjugates, 111–113
Condensation polymers, biomedical applications, 9–10
Cooley's anemia, iron chelating polymers and, 8–9
Copper:
   chelating polymers, biomedical applications, 7–9
   metallated DNA conjugates, 36–37
   metal-mediated basepairs:
      artificial nucleosides, 49–50
      self-assembled arrays, 53–54
      thermal stability, 51–52
Crosslinked compounds:
   biofissionable platinum-nitrogen complexes, 153–154
   DNA crosslinked cisplatin, 128–130
   glass ionomer technology, 196–198
Cyclic voltammetry:
   ferrocene-DNA conjugates, 30–34
   metallated DNA conjugates, 38–41
Cyclodextrins, platinum-oxygen-bound polymers, 166–180

Cyclopentadienyl, small-molecule analogs, biomedical applications, 11–16

D

Dendritic structures, platinum-oxygen-bound polymers, 165–180
Dental organic polyacid-inorganic composites:
   basic properties, 194
   glass ionomer technology, 194–202
      amino-acid-modified ionomers, 198–200
      N-vinylpyrrolidone (NVP)-modified ionomers, 201–206
Desferrioxamine B (DFO), biomedical applications, 8–9
Dextran derivatives, platinum-oxygen-bound polymers, 167–180
Dialkyltin compounds, anticancer activity, 59–62
cis-Diaminedichloroplatinum(II). *See* Cisplatin
Diamines, cisplatin mainchain structures, 138–141
1,2-Diamines, biofissionable platinum-nitrogen complex anchoring, 154–161
1,1-Dicarboxylatoplatinum complex, platinum-oxygen-bound polymers, 170–180
Dicarboxyl-containing compounds, platinum-oxygen-bound polymers, 162–180
Dichlorides, organotin compounds, anticancer activity, 59–62
Dichloro[rel-1*R*,2*S*]-1,2-cyclohexanediamine-N,N'-platinum(II) (DACH-Pt):
   mixed platinum-oxygen/platinum-nitrogen bound polymers, 181–182
   platinum-oxygen-bound polymers, 161–180
Dicyclohexylcarbodiimide (DCC), ester-linked ferrocene conjugates, 109–110
Diethyltin compounds, anticancer activity, 60–62
   polymers, 67–70
Differential pulse voltammography (DPV), ferrocene-DNA conjugates, 23–34
Differential scanning calorimetry (DSC), cisplatin mainchain thermal stability, 145
5'-Dimethoxytrityl-5-ethynylferrocene-2'-dexyuridine, synthetic pathways, 21–22
3-(Dimethylamino)propyl (DP), biofissionable platinum-nitrogen complex anchoring, 160–161
4-(Dimethylamino)pyridine (DMAP), ester-linked ferrocene conjugates, 109–110
Dimethylsulfoxide (DMSO):
   biofissionable platinum-nitrogen complexes, 151–154
   cisplatin mainchain structures, amine derivatives, 138–141
Dinuclear platinum complexes, polymeric platinum-containing anticancer compounds, 183

Diol characteristics, organotin polymer anticancer activity, 69–70

Diphenyltin compounds, anticancer activity, 60–62 polymers, 69–70

Diplatinum compounds, structure-activity relationships, 130–133

Divalent ions, metallated DNA, 37–41

DIVEMA compound, platinum-oxygen-bound polymers, 169–180

DNA crosslinking, cisplatin compounds, 128–130

DNA sequencing, ferrocene-DNA conjugates, 34

Double-stranded DNA (dsDNA):
electron transfer, 20
ferrocene-DNA conjugates, 26–34
metallated conjugates, 37–41

Drug conjugation:
ferrocene polymers, 98–100
platinum-oxygen-bound polymers, 163–180
polymeric ferrocene conjugates, 114–115
polymeric platinum-containing agents, 133–137

E

Ehrlich ascites tests:
ferrocene-ferricenium systems, 96–98
platinum-polyphosphazene complexes, 149

Electron transfer (ET):
double-stranded DNA, 20
ferrocene-ferricenium systems, 92–98
ferrocenes, 91
metal-labeled DNA:
Cu-DNA, 36–37
ferrocene-DNA conjugates, 22–34
ferrocene nucleotides, 20–22
M-DNA, 37–42
metallated DNA, 36–42
phenanthroline nucleosided, 36
ruthenium nucleosides, 34–36

Enhanced permeation and retention (EPR) effect:
ferrocene polymer-drug conjugation, 99–100
platinum-containing polymer-drug conjugation, 135

Enterbactin, biomedical applications, 8–9

Ester-linked ferrocene, polymer conjugates, 109–110

Extracellular species, cisplatin compounds, 128–130

F

Ferricenium:
bioactivity screening, 110–113
in biological environment, 92–98

Ferricenylalkylcarboxylates, biological environment, 93–98

Ferrichrome, biomedical applications, 8–9

Ferrocenes:
condensation polymers, 9–10
DNA conjugates, 22–34
metal-containing bioactive agents, 6
nucleotides, 20–22
polymeric conjugates:
amide-linked ferrocene conjugates, 102–109
basic properties, 90–91
bioactivity screening, 110–113
carrier components, 101–102
ester-linked conjugates, 109–110
ferrocene-ferricenium system, 92–98
future research issues, 113–114
pharmaceutical applications, 98–100
synthesis and structure, 100–110

Ferrocenylalkylated benzotriazoles, biological activity, 97–98

Ferrocenylalkylcarboxylates, biological environment, 93–98

Ferrocenylation agents, amide-linked ferrocene-polymer conjugates, 105–109

Ferrocenylthymidine derivatives, synthetic pathways, 21–22

Ferrocifens, small-molecule analogs, biomedical applications, 11–16

Ferroquine, biomedical applications, 13–16

Fourier transform infrared spectra (FTIR), N-vinylpyrrolidone (NVP)-modified glass ionomer synthesis, 203

Free radicals:
ferrocene-ferricenium systems, 92–98
ferrocenes, 91
in polymeric bioactive agents, 6

Fuji II control:
glass ionomer technology, 199–200
N-vinylpyrrolidone (NVP)-modified glass ionomers, 203–206

Fullerenes, small-molecule analogs, biomedical applications, 16

G

Germanium, condensation polymers, 9–10

Glass ionomers, organic polyacid-inorganic composites, 194–202
amino-acid-modified compounds, 198–200
N-vinylpyrrolidine (NVP)-modified compounds, 201–206

Glycine-phenylalanine-leucine-glycine (GFLG), mixed platinum-oxygen/platinum-nitrogen bound polymers, 180–182

Gold surfaces, ferrocene-DNA conjugates, 23–34

Guanine, cisplatin compounds, 128–130

H

Half-sandwich metal compounds, biomedical applications, 14–16

Halides, organotin compounds, anticancer activity, 59–62

Hard-soft theory, platinum(II) square-planar structure, 124–125

HBTU coupling agent, amide-linked ferrocene polymer conjugates, 102–109

HeLa cell line:
biofissionable platinum-nitrogen complexes, 152–154
cisplatin mainchain structures, amine derivatives, 138–141
platinum-oxygen-bound polymers, 173–180
polymeric ferrocene conjugates, 111–113

Himt. *See* Imidazoline-2(1,3*H*)-thione

Histidine, cisplatin mainchain polymers, 141–143

Hmimt. *See* 1-Methyl-imidazoline-2(3*H*)-thione

"Housekeeping" proteins:
organotin polymers, anticancer activity, 66–70
platinum-containing polymer-drug conjugation, 136

Howell-Walles patent, biofissionable platinum-nitrogen complex anchoring, 155–161

HTB161 cell lines, organotin polymer anticancer activity, 67–70

HTB 75 cells, organotin polymer anticancer activity, 67–70

Hydrazines, cisplatin mainchain structures, 143–144

Hydrochloric acid, organotin compounds, anticancer activity, 59–62

Hydrogen-bonded base pairing, alternative schemes, 46–48

Hydroxamic acid, iron chelating polymers, 9

Hydroxybenzosuccinimide (HOSu), ferrocene nucleotides, 20–22

Hydroxybenzotriazole (HOBt), ferrocene nucleotides, 20–22

2-Hydroxyethyl methacrylate (HEMA), glass ionomer technology, 196–198
amino acid ionomers, 200

Hydroxyl-containing compounds, platinum-oxygen-bound polymers, 162–180

Hydroxypropylmethacrylamide (HPMA):
biofissionable platinum-nitrogen complex anchoring, 160–161
mixed platinum-oxygen/platinum-nitrogen bound polymers, 180–182
platinum-oxygen-bound polymers, 165–180

Hydroxypyridone nucleoside, metal-mediated basepairs:
self-assembled arrays, 53–54
thermal stability, 52

I

IC_{50} value, biofissionable platinum-nitrogen complex anchoring, 160–161

IGROVE-1 cell line, platinum-oxygen-bound polymers, 164–180

Imidazoline-2(1,3*H*)-thione (Himt), organotin anticancer activity, 61–62

Immunosuppressive behavior:
ferrocene polymer-drug conjugation, 98–100
organotin compounds, 63–65

Immunotargeting studies, platinum-oxygen-bound polymers, 163–180

Intracellular species, cisplatin compounds, 128–130

Iron, chelating polymers, biomedical applications, 7–9

2-Isocyanatoethyl methacrylate (IEM), glass ionomer technology, 200

Isocytosine (iso-C), Watson-Crick basepairing, 47–48

Isoguanine (iso-G), Watson-Crick basepairing, 47–48

I-trans complexes, platinum-oxygen-bound polymers, 161–180

K

K562 cell line, organotin cytotoxicity, 63–65

KHN values, glass ionomer technology, 199–200

Kinetin hormone, polymer anticancer activity, 67–70

Klenow fragment I polymerase:
ferrocene-DNA conjugates, 25–34
non-hydrogen-bonding basepairs, 48–49

L

L929 cell lines, biofissionable platinum-nitrogen complexes, 152–154

Lead, biomedical applications, 3

Levamisole, cisplatin mainchain structures, 146

Lithium, biomedical applications, 15–16

LNCaP cell line, polymeric ferrocene conjugates, 111–113

Lysosomotropic compounds, polymeric ferrocene conjugates, 114–115

M

Macromolecular synthesis:
biofissionable platinum-nitrogen complex anchoring, 156–161
ferrocene polymer-drug conjugation, 99–100
organotin compound toxicity, 63–65
platinum-oxygen-bound polymers, 167–180

"Magic bullet" delivery systems, polymeric platinum-containing anticancer compounds, 183

Magnesium ions, metallated DNA, 37–41

Magnetic resonance imaging (MRI), small-molecule analogs, biomedical applications, 16

Mainchain structures:
cisplatin polymeric compounds, 137–147
amine derivatives, 137–141
amino acid derivatives, 141–143
antiviral activity, 145–147
nitrogen-platinum products, 143–144
solution stability, 144–145
thermal stability, 145
polymeric ferrocene conjugates, 114–115

Maleic acid (MA), *N*-vinylpyrrolidone (NVP)-modified glass ionomers, 202–203

MCF-7 cell line, organotin cytotoxicity, 63–65

Mercury:
biomedical applications, 3
metal-mediated basepairs, thermal stability, 51–52

Metal chelation polymers, biomedical applications, 7–9

Metallated DNA:
Cu-DNA, 36–37
divalent ions, 37–41

Metallocenes:
classification, 90–91
polymeric ferrocene conjugates, 113–114

Metal-mediated DNA basepairing:
artificial nucleosides, 49–50
future applications, 54
basic concept, 49
discrete self-assembled arrays, 52–54
thermal stability effects, 50–52

N-Methacryloylglutamic acid (MGA) monomer, glass ionomer technology, 198–200

Methicillin, organotin monomers, anticancer activity, 62–65

Methotrexate, cisplatin mainchain structures:
amine derivatives, 138–141
antiviral activity, 146–147
polymer degradation, 144–145

1-Methyl-imidazoline-2(3*H*)-thione (Hmimt), organotin anticancer activity, 61–62

Micelles, platinum-oxygen-bound polymers, 167–180

Michael addition polymerizations, platinum-oxygen-bound polymers, 171–180

Mitochondrial energy metabolism, organotin compound toxicity, 63–65

Mixed platinum-oxygen/platinum-nitrogen bound polymers, 180–182

Molecular weight, cisplatin mainchain solution stability, 144–145

Monomeric compounds, organotin, molecular-level studies, 62–65

3-(Morpholin-4-yl)propyl (MP), biofissionable platinum-nitrogen complex anchoring, 160–161

N

NADH, ferrocene-ferricenium systems, 92–98

Necrosis, organotin cytotoxicity, 64–65

Nickel ions:
metallated DNA, 37–41
metal-mediated basepairs, thermal stability, 51–52

Nitrates, platinum-oxygen-bound polymers, 177–180

Nitrogen compounds:
cisplatin mainchain structures:
amine derivatives, 137–141
miscellaneous products, 143–144
cisplatin structure-activity relationships, 130–133
platinum carrier-bound complexes, nitrogen donor ligands, 147–161
biofissionable complexes, amine anchors, 149–161
platinum-polyphosphazenes, 147–149
platinum(II) square-planar structure, 124–125
platinum-oxygen-bound polymers, 176–180

Non-hydrogen-bonding DNA base pairs, structure and characteristics, 48–49

Nuclear magnetic resonance, amide-linked ferrocene polymer conjugates, 104–109

Nucleophilicity, platinum(II) square-planar structure, 124–125

Nucleosides:
metal-mediated basepairs, 49–50
ruthenium-DNA conjugates, 34–36

N-vinylpyrrolidone (NVP)-modified glass ionomers:
basic properties, 201–202
characterization, 203
future applications, 204–206
materials for, 202–203
physical properties, 203
polymer synthesis, 203
recent modifications, 202–203

Nyquist plots, metallated DNA conjugates, 40–41

O

Octahedral compounds, cisplatin structure-activity relationships, 132–133

Oligonucleotides:
ferrocene-DNA conjugates, 22–34
metal-labeled DNA, ferrocene conjugates, 20–22
metal-mediated basepairs, 51–52
self-assembled arrays, 53–54

Organic polyacid-inorganic composites, dental applications:
  basic properties, 194
  glass ionomer technology, 194–202
    amino-acid-modified ionomers, 198–200
    N-vinylpyrrolidone (NVP)-modified ionomers, 201–206
Organometallics. *See also* specific compounds, e.g. Tin
  biomedical applications:
    metal-containing bioactive agents, 4–7
      ferrocene, polymeric systems, 6
      osmarins, 7
      tin-containing biocidal polymers, 5–6
      polymers, 7–10
        condensation polymers, 9–10
        metal chelation polymers, 7–9
      research background, 2–4
      small-molecule analogs, 11–16
  classification, 90–91
  square-planar structures, 121–125
Organotin:
  acyclovir agents in, 80–81
    bioactivity, 81–82
    experimental protocols, 82–83
    reactions and results, 83–86
  anticancer activity:
    chelating agents, 59–62
    monomeric molecular studies, 62–65
    polymer activity, 65–70
    research background, 58–59
  biocidal polymers, 5–6
  condensation polymers, 9–10
  mechanisms and reactions, 76–77
  structural characteristics, 77–80
Osmarins, polymeric toxicity moderation, 7
Oxamide, cisplatin mainchain structures, 143–144
Oxygen. *See also* Mixed platinum-oxygen/platinum-nitrogen bound polymers
  platinum-oxygen-bound polymers, 161–180

P

Palladium catalysts:
  ferrocene nucleotides, 21–22
  metal-mediated basepairs:
    artificial nucleosides, 49–50
    thermal stability, 51–52
  polymer synthesis, 125
Peptide nucleic acids (PNA), ferrocene-DNA conjugates, 34
Pharmacokinetics:
  cisplatin compounds, 125–126
    polymer-drug conjugation strategy, 133–137
  ferrocene polymer-drug conjugation, 98–100

platinum-oxygen-bound polymers, 164–180
polymeric ferrocene conjugates, 114–115
Phenanthroline nucleosides, DNA conjugates, 36
o-Phenylenediamine, biofissionable platinum-nitrogen complex anchoring, 155–161
pH levels:
  biofissionable platinum-nitrogen complexes, 153–154
  cisplatin mainchain polymers, amino acid derivatives, 142–143
  ferrocene-ferricenium systems, 97–98
Phosphoramidite derivatives:
  ferrocene-DNA conjugates, 22–23
  metal-mediated basepairs, 50–52
  ruthenium-DNA conjugates, 35–36
π-bonding, platinum(II) compounds, 123–125
Platinum. *See also* Mixed platinum-oxygen/platinum-nitrogen bound polymers
  condensation polymers, 9–10
  currently approved compounds, 125–126
  polymeric anticancer compounds:
    basic properties, 120–121
    carrier-bound complexes, nitrogen donor ligands, 147–161
      biofissionable complexes, amine anchors, 149–161
      platinum-polyphosphazenes, 147–149
    cisplatin properties, 127–130
    conjugation strategy, 133–137
    formation mechanisms, 121–125
    future applications, 182–184
    mainchain-incorporated *cis*-diamine-coordinated platinum, 137–147
      amine derivatives, 137–141
      amino acid derivatives, 141–143
      antiviral activity, 145–147
      nitrogen-platinum products, 143–144
      solution stability, 144–145
      thermal stability, 145
    mixed platinum-oxygen/platinum-nitrogen-bound polymers, 180–182
    monomeric structures, 125–126
    nomenclature, 125
    platinum-oxygen-bound polymers, 161–180
    structure-activity relationships, 130–133
  small-molecule analogs, biomedical applications, 12–16
Poly(AA-*co*-IA-*co*-MGA), glass ionomer technology, 199–200
Poly(AA-*co*-IA-*co*-NVP):
  basic properties, 201–202
  compositions and viscosities, 205
  glass ionomer experiments, 204–206

Poly(acrylic acid-*co*-maleic acid), glass ionomer technology, 196–198
Polyamidoamine Starburst (PAMAM) dendrimer, platinum-oxygen-bound polymers, 165–180
Polyaspartamides:
    amide-linked ferrocene polymer conjugates, 102–109
    as antiproliferative agents, 102
    biofissionable platinum-nitrogen complex anchoring, 156–161
Polyethyleneimine (PEI), biofissionable platinum-nitrogen complexes, 151–154
Poly-(L-lysine), platinum-oxygen-bound polymers, 168–180
Polymers:
    biomedical applications, 7–10
        condensation polymers, 9–10
        metal chelation polymers, 7–9
    ferrocene conjugates:
        amide-linked ferrocenes, 102–109
        basic properties, 90–91
        bioactivity screening, 110–113
        carrier components, 101–102
        ester-linked conjugates, 109–110
        ferrocene-ferricenium system, 92–98
        future research issues, 113–114
        pharmaceutical applications, 98–100
        synthesis and structure, 100–110
    metal-containing bioactive agents, 4–7
        ferrocenes, 6
        osmarins, 7
        tin-containing biocides, 5–6
    N-vinylpyrrolidone (NVP)-modified glass ionomer synthesis, 203
    organotin compounds:
        anticancer activity, 65–70
        biocides, 5–6
        research background, 58
    platinum-containing anticancer compounds:
        basic properties, 120–121
        carrier-bound complexes, nitrogen donor ligands, 147–161
        biofissionable complexes, amine anchors, 149–161
        platinum-polyphosphazenes, 147–149
        cisplatin properties, 127–130
        conjugation strategy, 133–137
        formation mechanisms, 121–125
        future applications, 182–184
        mainchain-incorporated *cis*-diamine-coordinated platinum, 137–147
        amine derivatives, 137–141
        amino acid derivatives, 141–143
        antiviral activity, 145–147

nitrogen-platinum products, 143–144
        solution stability, 144–145
        thermal stability, 145
    mixed platinum-oxygen/platinum-nitrogen-bound polymers, 180–182
    monomeric structures, 125–126
    nomenclature, 125
    platinum-oxygen-bound polymers, 161–180
    structure-activity relationships, 130–133
Polyphosphazenes, platinum polymer-bound complexes, 147–149
Polysaccharides, platinum-oxygen-bound polymers, 179–180
Polyvinylamide, biofissionable platinum-nitrogen complexes, 152–154
Polyvinylpyrrolidone (PVP), biofissionable platinum-nitrogen complex anchoring, 155–161
Potassium persulfate, *N*-vinylpyrrolidone (NVP)-modified glass ionomers, 202–203
Prodrug design:
    ferrocene polymer-drug conjugation, 99–100
    platinum-containing polymer-drug conjugation, 135
Protein delivery agents, polymeric platinum-containing anticancer compounds, 183–184
Proton nuclear magnetic resonance, amide-linked ferrocene polymer conjugates, 104–109
Pseudo-first-order conditions, platinum square-planar structure, 122–125
Pyran-5,6-diyl derivatives, platinum-oxygen-bound polymers, 169–180

R
Reactive oxygen species (ROS), organotin cytotoxicity, 63–65
"Reagentless" detection, ferrocene-DNA conjugates, 28–34
Redox-active DNA, ferrocene-DNA conjugates, 26–34
Redox behavior, ferrocene-ferricenium systems, 92–98
Reduction potentials, ferrocenylation agents, 105–109
Ruthenium:
    DNA conjugates, 34–36
    small-molecule analogs, biomedical applications, 12–16
Rutilus rubilio, organotin monomers, anticancer activity, 63–65

S
Salt bridges, glass ionomer technology, 194–198
Scavenger systems, ferrocene-ferricenium systems, 93–98

Self-assembled monolayers (SAMs):
  electron transfer, 20
  ferrocene-DNA conjugates, 23–34
  metal-mediated basepairs, 52–54
Side-chain structures:
  amide-linked ferrocene-polymer conjugates, 107–109
  biofissionable platinum-nitrogen complexes, 150–154
  polymeric ferrocene conjugates, 111–113
Signaling probes, ferrocene-DNA conjugates, 29–34
σ-bonding, platinum(II) compounds, 123–125
Silver, metal-mediated basepairs:
  artificial nucleosides, 49–50
  thermal stability, 51–52
Single-strand DNA (ssDNA), ferrocene-DNA conjugates, 28–34
Size effects, cisplatin mainchain polymers, amine derivatives, 140–141
Small-molecule analogs, biomedical applications, 11–16
Solubility parameters:
  biofissionable platinum-nitrogen complexes, 153–154
  N-vinylpyrrolidone (NVP)-modified glass ionomers, 201–202
  platinum-oxygen-bound polymers, 177–180
  polymeric ferrocene conjugates, 114–115
Solution stability, cisplatin mainchain structures, 144–145
Spacer link functionality, glass ionomer technology, 198–200
Square-planar structure:
  biofissionable platinum-nitrogen complexes, 149–154
  cisplatin structure-activity relationships, 132–133
  platinum(II) complex formation, 121–125
Stem loop structure, ferrocene-DNA conjugates, 28–34
Stepwise aminolytic treatment, biofissionable platinum-nitrogen complex anchoring, 156–161
Structure-activity relationships, cisplatin compounds, 130–133
Superoxide dismutase (SOD), ferrocene-ferricenium systems, 95–98
Superoxide radicals, ferrocene-ferricenium systems, 93–98

T

Tartaric acid, glass ionomer technology, 194–198
Technetium compounds, organotin anticancer activity, 59–62
Technetium-99m sestamibi, biomedical applications, 14–16

Tert-amine side groups, amide-linked ferrocene-polymer conjugates, 106–109
Tetrachloropalatinate:
  biofissionable platinum-nitrogen complex anchoring, 157–161
  platinum-oxygen-bound polymers, 165–180
Tetrahaloplatinate(II), platinum(II) square-planar structure, 124–125
Tetramisole, cisplatin mainchain structures, 146
Thermal gravimetric analysis (TGA), cisplatin mainchain thermal stability, 145
Thermal stability:
  cisplatin mainchain structures, 145
  metal-mediated basepairs, 50–52
Thiophosphate groups, ferrocene-DNA conjugates, 24–34
Thiourea, cisplatin mainchain structures, 143–144
Thymine, non-hydrogen-bonding basepairs, 48–49
Thymocytes, tin toxicity, 58
Thymus, organotin anticancer activity, 61–62
Tin. *See* Organotin
Toxicity:
  cisplatin compounds, 129–130
  organometallics, 2–4
  osmarin, polymeric moderation, 7
  platinum-oxygen-bound polymers, 166–180
  small-molecule analogs, biomedical applications, 13–16
  tin compounds, 58
    anticancer activity, 62–65
Trans effect order, platinum square-planar structure, 122–125
Trans influence:
  cisplatin structure-activity relationships, 130–133
  platinum square-planar structure, 123–125
Triethyltin acetate, toxicity, 58
Triethyltin lupinylsulfide hydrochloride, organotin cytotoxicity, 64–65
Tungsten, small-molecule analogs, biomedical applications, 12–16

V

Vanadium, small-molecule analogs, biomedical applications, 12–16
Vinylsulfonate, biofissionable platinum-nitrogen complexes, 152–154
Viral DNA, ferrocene-DNA conjugates, 25–34
Visible-light-curable (VLC) formulation:
  glass ionomer technology, 194–198
  amino acid ionomers, 200
  N-vinylpyrrolidone (NVP)-modified glass ionomers, 202
Voltage-gated potassium currents, organotin compound toxicity, 63–65

W

Water solubility, polymer-ferrocene conjugates, 101–102
Watson-Crick basepairing:
  alternative hydrogen-bonds, 46–48
  non-hydrogen-bonding basepairs, 48–49

Y

Yoshida ascites tests, ferrocene-ferricenium systems, 96–98

Z

Zinc ions, metallated DNA, 37–41